FATIGUE LIFE PREDICTION OF SOLDER JOINTS IN ELECTRONIC PACKAGES WITH ANSYS®

THE KLUWER INTERNATIONAL SERIES
IN ENGINEERING AND COMPUTER SCIENCE

FATIGUE LIFE PREDICTION OF SOLDER JOINTS IN ELECTRONIC PACKAGES WITH ANSYS®

by

Erdogan Madenci
Ibrahim Guven
Bahattin Kilic
The University of Arizona
Tucson, AZ, U.S.A.

KLUWER ACADEMIC PUBLISHERS
Boston / Dordrecht / London

Distributors for North, Central and South America:
Kluwer Academic Publishers
101 Philip Drive
Assinippi Park
Norwell, Massachusetts 02061 USA
Telephone (781) 871-6600
Fax (781) 681-9045
E-Mail < kluwer@wkap.com>
E-Mail < services@wkap.nl>

Distributors for all other countries:
Kluwer Academic Publishers Group
Post Office Box 322
3300 AH Dordrecht, THE NETHERLANDS
Telephone 31 78 6576 000
Fax 31 78 6576 254

 Electronic Services < http://www.wkap.nl>

Library of Congress Cataloging-in-Publication Data

Madenci, Erdogan.
  Fatigue life prediction of solder joints in electronic packages with ANSYS® / Erdogan Madenci, Ibrahim Guven, Bahattin Kilic.
  Boston, Mass. : Kluwer Academic Publishers, 2002.
  p. cm.—(The Kluwer international series in engineering and computer science; SECS 719)
  Includes bibliographical references and index.

  Additional material to this book can be downloaded from http://extras.springer.com.

  ISBN: 1-4020-7330-5
    1. Electronic packaging—Computer-aided design. 2. Solder and soldering—Quality control. 3. Metals—fatigue. 4. Service life (Engineering)—forecasting. 5. ANSYS® (computer system).

  TK7870.15 .M32 2002
  621.381'046—dc21

                                                        2002038928

ANSYS® is a registered trademark of ANSYS, Inc.

Printed on acid-free paper.

Printed in the United States of America.

PREFACE

The fatigue life prediction of solder joints is an important issue in the electronics industry. Over the last decade, one method has emerged as the most widely used for a multitude of package configurations. However, this method requires knowledge of finite element modeling and simulation with ANSYS®, a commercially available finite element program. Furthermore, the three-dimensional finite element modeling of any electronic package remains a formidable task, even if the analyst has extensive knowledge of ANSYS®. This book describes the method in great detail, starting from its theoretical basis. The reader is supplied with a group of ANSYS® macros, ReliANS, that are integrated into the ANSYS®-GUI. These macros have been specifically designed for the life prediction analysis of solder joint fatigue in electronic packages. ReliANS transforms the complicated fatigue life prediction analysis process into a sequence of simple building blocks. In order to avoid any confusion, examples are utilized to explain specific steps of the analysis method. Case studies are included to assist the readers in regenerating the solutions on their computers.

Upon installation of ReliANS, a menu item named "Packaging" is added to the preprocessor, solution, and postprocessor menus within the ANSYS®-GUI environment. The packaging menu under the preprocessor menu concerns model generation and material properties. Under the solution menu, the packaging menu contains items related to the application of boundary conditions and the simulation of the thermal or mechanical cycle. The post-processor menu is used for calculating the expected life of the critical solder joint under consideration.

Chapter 1 provides an overview of several mathematical modeling techniques for inelastic solder behavior and several solder joint failure prediction methods. In Chapter 2, the methods utilizing the finite element method with ANSYS for the thermomechanical fatigue life prediction analysis of electronic packages are presented. Following the theoretical basis, a case study is considered in detail. The mechanical fatigue life prediction analysis of electronic packages is presented in Chapter 3, also with a case study. A compilation of the add-on ANSYS® macros, ReliANS, along with detailed information for each of the macros, is presented in Chapter 4. The installlation instructions for ReliANS are included in Appendix A, and the command line input listings for the two case studies in Chapters 2 and 3 are given in Appendix B.

We would like to acknowledge the support and information provided by the Semiconductor Research Corporation (SRC), its member companies, and Ericsson Components AB of Sweden.

We are greatly indebted to Connie Spencer for her invaluable efforts in typing, editing, and assisting with each detail associated with the completion of this book.

TABLE OF CONTENTS

LIST OF COMMANDS

LIST OF TABLES

LIST OF FIGURES

Chapter 1

INTRODUCTION

The degree of complexity of an electronic package is identified by three distinct assembly levels: (1) attachment of the chip (die) to the chip carrier (substrate), (2) attachment of the chip carrier and the die to the printed circuit board (PCB), and (3) the attachment of the PCB to the other PCBs through edge connects or to other devices through cables. The Surface Mount Technology (SMT), commonly used in the electronics industry, enables the attachment of the chip directly to the PCB. The flip-chip technology, an advanced form of surface mount technology, for attaching a chip equipped with solder bumps on its surface to the substrate and then to the PCB with a ball grid array connection is described in Fig. 1.1.

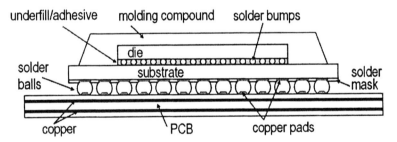

Fig. 1.1   Flip-chip components.

As the performance requirements of an electronic package increase, its size decreases while requiring fast heat dissipation during its operation. One major concern of flip-chip technology is the failure of solder joint interconnections due to large temperature excursions in the range of -25 to 100°C. The mismatch in the coefficients of thermal expansion (CTE) between the PCB and the components within the chip carrier (substrate), including the die, results in the thermal fatigue failure of the interconnection solder joints. Under such thermal cycling conditions, the solder joint experiences elastic, plastic, and creep deformations, leading eventually to cracking and separation.

The reliability of an electronic package is dictated by its electrical, thermal, and mechanical performances. In order to enhance its mechanical performance, a numerical simulation of the mechanical behavior of an electronic package is essential. However, the fidelity of the simulation is dependent upon accurate numerical modeling, the material models, and the failure criterion. Failure mechanisms in an electronic package can be classified as intrinsic and extrinsic. The intrinsic failures arise from the process-related defects in silicon wafer substrates, dielectrics, insulating films, interconnecting films between devices, and components. The extrinsic failures arising from the external mechanical and thermal loads, and environmental conditions, may include chip fracture, solder joint fracture, moisture-induced swelling and cracking of the encapsulation, corrosion, creep, and fatigue of solder joints.

1.1 Numerical Modeling with Finite Element Analysis

An accurate numerical modeling of the stress and strain fields in electronic packages is imperative for the improvement of current and future mechanical designs of electronic packages. The most commonly accepted models employing the finite element method for thermal fatigue life prediction analyses can be grouped as: (1) nonlinear slice model, (2) global model with linear super elements and nonlinear solder, (3) linear global model with a nonlinear submodel, (4) nonlinear global model with a nonlinear submodel, and (5) nonlinear global model. In the global analyses with a submodel, the displacement fields obtained from the analyses are extrapolated for different temperatures and applied as boundary conditions in the submodel of the solder joint with a refined mesh. In the nonlinear analyses, four thermal cycles are sufficient to achieve a steady-state response (stable hysteresis loop). According to the comparative analysis by Gustafsson et al. (2000), a nonlinear global model, with a rather coarse mesh and a subsequent nonlinear submodeling of the critical solder joint with a refined mesh, has more fidelity than the other models.

1.2 Constitutive Relations

In establishing the fidelity of a numerical model, a precise description of the material behavior for each constituent of the electronic package is extremely critical. The components of an electronic package may exhibit temperature- and time-dependent material behavior. Also, their material behavior at the length scales characteristic of an electronic package is usually different from that of the bulk properties. In addition, the solder behavior is dependent on strain rate, stress, and temperature because of its high homologous temperature. The material property data for typical components of an electronic package are provided in Tables 1.1 (temperature-dependent) and 1.2 (temperature-independent). The parameters E, v, and α represent the Young's modulus, Poisson's ratio, and coefficient of thermal expansion for an isotropic material. For an orthotropic material, the parameters E_i and α_i represent the properties along the ith direction, with $i = x, y, z$.

The deformation behavior of solder is dependent on temperature and time. Therefore, the deformation of solder is described by time-dependent constitutive laws of viscoplasticity or time-dependent creep combined with time-independent plasticity. In viscoplastic deformation, the elastic region is bounded by a so-called static yield surface in stress space, and all inelastic deformations are time-dependent. Also, the inelastic deformations occur at all non-zero stress values. The viscoplastic constitutive law does not distinguish the plastic strains from those of creep. Unlike the time-independent

Table 1.1   Material properties in a typical electronic package–
temperature dependent.

		Temperature (°K)					
		240	280	320	360	400	440
SnPb (63/37)	E (MPa)	39435	33367	27299	21231	15163	9095
Copper	α (ppm/°C)	16.066	16.443	16.821	17.198	17.576	17.954
Si	α (ppm/°C)	2.0154	2.4188	2.7160	2.9651	3.2240	3.5508

Table 1.2   Material properties in a typical electronic package–
temperature independent.

	E_x, E_z (MPa)	E_y (MPa)	v_{xz} (ppm/°C)	v_{xy}, v_{yz} (ppm/°C)	α_x, α_z	α_y
SnPb (63/37)			0.35		24.5	
Copper	128930		0.344			
Si	162700		0.278			
BT[a] – 1	5170		0.33		160	
BT[a] – 2	16892	7377	0.11	0.39	15	57
AlN	331×10^3		0.23		4.5	
Al$_2$O$_3$	345×10^3		0.23		7.1	
Epoxy	2410		0.3		60	
PCB – FR4	20000	10000	0.38		18	50
Molding compound	11300		0.3		14	
Die attach	7400		0.4		52	
Solder mask	2000		0.3		70	

[a]Bismaleimide triazine.

plasticity law, the viscoplastic constitutive law does not rely on an explicit yield surface and the loading and unloading criterion. Instead, it utilizes an internal state variable, representing the resistance of the material to inelastic deformations. Creep deformation is time and temperature dependent, and the time-independent plastic deformation results in plastic strains depending on the yield surface and loading and unloading criterion.

There is no commonly accepted constitutive law for describing the behavior of solder. Furthermore, many of the sources of material data differ from one another. Among the various time-dependent constitutive laws for solder joints in electronic packages, the viscoplastic constitutive law introduced by Anand (1982) is frequently used. This material model is available in the ANSYS® material library (referred to as Anand's viscoplastic model).

Anand's model consists of two coupled differential equations that relate the inelastic strain rate to the rate of deformation resistance. The strain rate equation is

$$\frac{d\varepsilon_{in}}{dt} = A\left[\sinh\left(\xi\frac{\sigma}{s}\right)\right]^{(1/m)} e^{-Q/RT} \tag{1.1}$$

and the rate of deformation resistance equation is

$$\dot{s} = \left\{h_0\left(|B|\right)^a \frac{B}{|B|}\right\}\frac{d\varepsilon_{in}}{dt} \tag{1.2}$$

where

$$B = 1 - \frac{s}{s^*} \quad \text{and} \quad s^* = \hat{s}\left[\frac{1}{A}\frac{d\varepsilon_{in}}{dt}e^{-Q/RT}\right]^n \tag{1.3}$$

The numerical values and the definitions of these parameters for a eutectic solder alloy reported by Darveaux (1996) and Liu et al. (2002) are given in Table 1.3. It is worth noting that the initial value of deformation resistance (S_0), although not appearing in the governing equations, is required for the solution of these equations.

An alternative to the viscoplastic constitutive law is the combined time-dependent creep and time-independent plasticity models. The inelastic strain is composed of the plastic strains, ε_p, and creep strains, ε_c, as

$$\varepsilon_{in} = \varepsilon_p + \varepsilon_c \tag{1.4}$$

Table 1.3   Numerical values for a eutectic solder alloy in Anand's material model.

Parameter	Description	Value
$d\varepsilon_{in}/dt$	Effective inelastic strain rate	N/A
σ	Effective true stress	N/A
s	Deformation resistance	N/A
T	Absolute temperature (°K)	N/A
S_0	Initial deformation resistance	1800 (psi), 12.41 (MPa)
A (1/sec)	Pre-exponential factor	$(4 \times 10^6)^*$ $(3.785 \times 10^7)^{**}$
ξ	Stress multiplier	$(1.5)^*$ $(5.91)^{**}$
m	Strain rate sensitivity of stress	$(0.303)^*$ $(0.143)^{**}$
Q/R	Ratio of activation energy to universal gas constant (1°/K)	$(9400)^*$ $(11945)^{**}$
h_0	Hardening/softening constant	2×10^5 (psi), 1379 (MPa)
$\hat{s}$	Coefficient for deformation resistance saturation value	2×10^3 (psi), 13.79 (MPa)
n	Strain rate sensitivity of saturation value	0.07
a	Strain rate sensitivity of hardening or softening	1.3

*Reported by Darveaux (1996).
** Reported by Liu et al. (2002).

The creep strain rate is decomposed in terms of the transient, $\dot{\varepsilon}_{ct}$, and steady-state, $\dot{\varepsilon}_{ss}$, components as

$$\frac{d\varepsilon_c}{dt} = \frac{d\varepsilon_{ss}}{dt} + \frac{d\varepsilon_{ct}}{dt} \tag{1.5}$$

in which the steady-state component is the dominant part of the creep deformation. The total creep strain rate or its steady-state component can be described by either a power law or a sinh law of the form

$$\frac{d\varepsilon_c}{dt} = C_1 \sigma^{C_2} \varepsilon^{C_3} e^{-C_4/T} \tag{1.6}$$

in which C_i, $i = 1, 2, 3, 4$, are independent of temperature. This creep law is available in the ANSYS® material library. A modified form of this relation with time-dependent coefficients C_1 and C_3 was introduced by Mukherjee et al. (1969) and later modified by Lam et al. (1979) as

$$\frac{d\varepsilon_{ss}}{dt} = C_1(T)\sigma^{C_2} e^{C_3(T)} \tag{1.7}$$

Akay et al. (1997) utilized this relation in modeling the deformation of solder joints in terms of specific parameters in the form

$$\frac{d\varepsilon_{ss}}{dt} = \frac{AGb}{kT}\left(\frac{b}{d}\right)^P \left(\frac{\sigma}{G}\right)^n D_0\, e^{(-Q/RT)} \qquad (1.8)$$

The specific values for the constants reported by Akay et al. (1997) are given in Table 1.4.

Table 1.4   Material constants for power-law creep as employed by Akay et al. (1997).

Parameter	Description	Value
σ	Applied effective stress	N/A
G	Shear modulus	$2.2\times10^4 - 16.1\,T$ (MPa)
B	Burger's vector magnitude	3.2×10^{-7} (mm)
K	Boltzmann's constant	1.38×10^{-23} (J/°K)
T	Absolute temperature	N/A
D	Grain size	5.5×10^{-3} (mm)
D_0	Pre-exponential constant	100 (mm^2)
Q	Activation energy	44 (kJ/mole)
R	Gas constant	8.314 (J/°K mole)
N	Stress exponent	2.4
P	Grain size exponent	1.6
A	Dimensionless constant	40

The extension of the simple power-law relation to a double power-law relation was introduced by Knecht and Fox (1990) in the form

$$\frac{d\varepsilon_c}{dt} = A\sigma^C e^{\frac{-B}{T}} + D\sigma^F e^{\frac{-E}{T}} \qquad (1.9)$$

where σ is the effective stress, T is temperature, and A, B, C, D, E, and F are material constants. A detailed discussion of this model, commonly referred to as the "Integrated Matrix Creep Model," and the numerical values of the constants are provided by Iannuzzelli et al. (1996) as given in Table 1.5.

As introduced by Darveaux et al. (1995), a sinh-law creep model can be constructed in the form

$$\frac{d\varepsilon_c}{dt} = \frac{d\varepsilon_{ss}}{dt}\left(1 + \varepsilon_T B_T e^{-B_T \dot{\varepsilon}_{ss} t}\right) \qquad (1.10)$$

with

$$\frac{d\varepsilon_{ss}}{dt} = C_{St}\left[\sinh\left(\alpha_{1t}\sigma\right)\right]^n e^{-Q_a/kT} \qquad (1.11)$$

The numerical values of the constants are given in Table 1.6.

Table 1.5   Material constants in the Integrated Matrix
Creep Model (Iannuzzelli et al. 1996).

Parameter	Stress Measured in Units of	
	Psi	MPa
A	0.762×10^{-4}	1.6
B	10428	5794
C	2.0	2.0
D	5.95×10^{-18}	1.3×10^{-3}
E	17519	9733
F	7.1	7.1

Table 1.6   Material constants in Eqs. (1.10) and
(1.11) (Darveaux et al. 1995).

Parameter	Value
C_{5t}	8.03×10^4 (1/sec)
α_{1t}	4.62×10^{-4} (1/psi)
n	3.3
Q_a	0.7 eV
k	1.38×10^{-23}
T	Absolute temperature (°K)
ε_T	0.023
B_t	263

Another form of the sinh-law creep model can be constructed by negating the effects of deformation resistance in Anand's viscoplastic model, thus reducing Eq. (1.1) to the form

$$\frac{d\varepsilon_c}{dt} = A\left[\sinh\left(\xi\sigma\right)\right]^{(1/m)} e^{\left(-Q_a/kT\right)} \qquad (1.12)$$

It is worth noting that the ANSYS® software has several different implicit creep models readily available.

The time-independent plastic strain of solder can be described by the Ramberg-Osgood material model with isotropic and/or kinematic hardening. As an alternative to the multi-linear material models with isotropic hardening available in the ANSYS® material library, the time-independent plastic behavior of solder can be represented by a strain hardening law suggested by Syed (2001) of the form

$$\varepsilon_p = C_{6t}\left(\frac{\sigma}{G}\right)^m \qquad (1.13)$$

The numerical values of the constants, C_{6t} and m, and the shear modulus, G, are given by Darveaux et al. (1995) as $C_{6t} = 3.35 \times 10^{11}$, $m = 5.53$, and $G = G_0 - G_1(T - 273)$, with $G_0 = 1.9 \times 10^6$ psi and $G_1 = 8,100$ psi/°K. Under a room temperature of $T = 298$°K, the shear modulus becomes $G = 1.697 \times 10^6$ psi.

An examination of this strain hardening law reveals that plastic deformation starts as soon as a non-zero stress state develops. Thus, the yield surface is a point and translates without any expansion. This phenomenon, referred to as kinematic strain hardening, is valid only during the first continuous loading. Therefore, this strain hardening law is modified to include the effect of loading, unloading, and reloading. The modified form of this law for unloading and reloading can be expressed as

$$\varepsilon_p = \tilde{\varepsilon}_p - \frac{C_{6t}}{G^m} 2^{1-m} (\tilde{\sigma} - \sigma)^m \qquad \text{unloading} \qquad (1.14)$$

$$\varepsilon_p = \bar{\varepsilon}_p + \frac{C_{6t}}{G^m} 2^{1-m} (\sigma - \bar{\sigma})^m \qquad \text{reloading} \qquad (1.15)$$

where $\bar{\sigma}$ and $\bar{\varepsilon}_p$ are the known stress and strain values at the end of each load cycle and $\tilde{\sigma}$ and $\tilde{\varepsilon}_p$ are the known stress and strain values at the middle of each load cycle, as shown in Fig. 1.2.

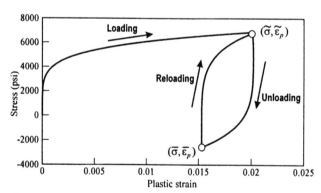

Fig. 1.2   Stress versus plastic strain behavior governed by Eqs. (1.13)-(1.15).

1.3 Failure Prediction

There exist many thermal fatigue life prediction models for determining the solder joint reliability of electronic packages, each with its own merits. Their salient features and a comparison among them are given by Lee et al.

(2000). These models are dependent on the package-level statistical (empirical) failure parameters associated with a key parameter of the structure, such as viscoplastic strain energy or matrix creep strain calculated using finite element analysis. The leading failure indicators for the correlation of thermal fatigue life are based on strain ranges such as the total, inelastic, and matrix creep strains (Iannuzzelli et al. 1996, Syed 1997) and the viscoplastic strain energy density increment by Darveaux (1996).

Although acceptable for determination of the package-level phenomenon of the number of thermal cycles to crack initiation, these failure prediction models do not account for the presence of a crack while predicting the number of thermal cycles to failure after crack initiation. They are based on the assumption of a *constant* crack growth rate. However, the presence of a crack introduces a variable crack growth rate, as measured by Lau et al. (2000) in their experimental investigation. Thus, the assumption of a *constant* crack growth rate after crack initiation may lead to erroneous life-cycle predictions.

These fatigue models provide the ability for comparative modeling (indexing) of new packages via existing qualified electronic packages. Also, they provide acceptable options for the material and geometry while identifying potential areas of reliability concern. As shown by Anderson et al. (2000), these predictive models do not consistently predict the life of a solder joint because the empirical parameters are dependent on the geometry, loading, and finite element discretization of the particular package types. Therefore, they cannot effectively account for varying package configurations. Depending on the package type, they either overestimate or underestimate the measured characteristic life. It is not enough to predict life based solely on the empirical parameters. The life predition models should include not only empirical parameters, but also a fracture parameter intrinsic to solder joints. This requires an accurate measurement of the critical fracture parameter, such as the critical strain energy density, as well as the empirical constants. Because of the process-dependent nature of the solder joint and the change in length scale, the critical fracture parameter for the solder joint cannot be measured from the bulk of the solder material. However, it can be extracted through an inverse approach by utilizing crack growth measurements in conjunction with finite element modeling and simulation.

One of the widely accepted failure criteria introduced by Darveaux (1997) for thermal fatigue life correlation is based on a relationship in terms of the volume-weighted-average inelastic work density increment, $\Delta \overline{W}_i$, the number of cycles to crack initiation, N_0, and the crack propagation rate, da/dN. Its explicit form is given by

$$N_0 = K_1 \Delta \bar{W}_i^{K_2} \tag{1.16}$$

$$\frac{da}{dN} = K_3 \Delta \bar{W}_i^{K_4} \tag{1.17}$$

where K_1, K_2, K_3, and K_4 are the empirical constants and a is the characteristic crack length (solder joint diameter at the crack site). These parameters were determined by Darveaux (2000) through curve-fitting against measurements in conjunction with the finite element simulations using the ANSYS® software. Invoking a *constant* crack propagation rate, the characteristic life, N_α, is calculated as

$$N_\alpha = N_0 + \frac{a}{da/dN} \tag{1.18}$$

The parameter $\Delta \bar{W}_i$, representing the volume-weighted-average inelastic work density accumulated per thermal cycle, is defined as

$$\Delta \bar{W}_i = \left(\sum_{n=1}^{\text{\# of elements}} \Delta W^{(n)} \times V_n \right) \Big/ \left(\sum_{n=1}^{\text{\# of elements}} V_n \right) \tag{1.19}$$

where $\Delta W^{(n)}$ designates the inelastic work density in the nth element, whose volume is denoted by V_n. Because the value of $\Delta W^{(n)}$ is dependent on the thickness of the finite elements, the empirical parameters K_1, K_2, K_3, and K_4 are intrinsically dependent on geometry, loading, and modeling assumptions within the scope of the finite element analysis. As reported by Gustafsson et al. (2000), the type of finite element model influences the inelastic work density calculation significantly.

The values of the empirical constants K_1, K_2, K_3, and K_4 for thermal cyclic loading, provided by Darveaux (2000) are given in Table 1.7.

Table 1.7　　Numerical values of empirical constants used in thermal fatigue life prediction (Darveaux 2000).

K_1	71000 cycles/psi^{K_2}
K_2	-1.62
K_3	2.76×10^{-7} in./cycle/psi^{K_4}
K_4	1.05

In the case of mechanical bending fatigue, the numerical values for the empirical constants, K_i, differ from those of thermal fatigue. Their numerical values as reported by Darveaux and Syed (2000) are given in Table 1.8.

Table 1.8   Numerical values of empirical constants used in mechanical fatigue life prediction (Darveaux and Syed 2000).

K_1	$111,000$ cycles/psi^{K_2}
K_2	-1.62
K_3	0.951×10^{-7} in./cycle/psi^{K_4}
K_4	1.21

An alternative to the existing models that invoke the assumption of a *constant* crack growth rate, Guven et al. (2001) employed the strain energy density criterion introduced by Sih (1973) in conjunction with three-dimensional thermo-mechanical finite element analysis to determine the life prediction of electronic packages. The three-dimensional finite element analysis and simulation include the appropriate loading path and regions of singular stress fields (crack fronts).

The number of thermal cycles to failure is associated with crack initiation and crack growth. The number of thermal cycles to crack initiation is influenced by the overall package geometry, material system, and thermal loading. Therefore, it can only be predicted by an empirical relationship similar to that of Pan (1994) in the form

$$N_0 = \frac{(dW/dV)^0_{cr}}{A\overline{W}_p + B\overline{W}_c} \qquad (1.20)$$

where $\overline{W}_p$ and $\overline{W}_c$ are the volume-weighted-average plastic and creep strain energy densities per cycle. The parameter $(dW/dV)^0_{cr}$ represents the critical strain energy density for crack initiation. Its value and the empirical constants A and B are obtained through curve-fitting to life-cycle measurements. In the case of thermal cyclic loading, their numerical values were determined by Pan (1994) by utilizing the experimental measurements by Hall and Sherry (1986). Their numerical values are $(dW/dV)^0_{cr} = 455 \times 10^{-9}$ J/mm^3 per solder volume, $A = 1$, and $B = 0.13$.

The number of thermal cycles after crack initiation can be determined by applying the strain energy density criterion introduced by Sih (1973) and Sih and MacDonald (1974), and later applied to fatigue crack growth by Sih and Moyer (1983). The underlying assumptions and extensive applications of this criterion are presented by Sih (1991).

Based on the study by Sih and Moyer (1983), crack growth under cyclic loading occurs when the strain energy density of the material along the crack front exceeds the critical strain energy density, $(dW/dV)_{cr}$. This assumption can be interpreted mathematically as

$$\left(\frac{dW}{dV}\right)_{cr} = \Delta\left(\frac{dW}{dV}\right)\Delta N + \left(\frac{dW}{dV}\right)_{avg} \tag{1.21}$$

where $\Delta(dW/dV)$ is the strain energy density increment per thermal cycle during a number of cycles, ΔN. The average strain energy density, $(dW/dV)_{avg}$, at a distance Δr ahead of the crack front is defined by

$$\left(\frac{dW}{dV}\right)_{avg} = \frac{1}{2}\left\{\frac{1}{2}\left[\left(\frac{dW}{dV}\right)_{t_f} + \left(\frac{dW}{dV}\right)_{t_i}\right] + \left(\frac{dW}{dV}\right)_{t_m}\right\} \tag{1.22}$$

in which $(dW/dV)_{t_f}$, $(dW/dV)_{t_i}$, and $(dW/dV)_{t_m}$ represent the strain energy density values at the end (t_f), start (t_i), and middle (t_m) of the thermal cycle, respectively. A typical thermal cycle with definitions of t_f, t_i, and t_m is described in Fig. 1.3.

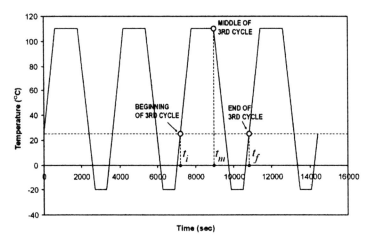

Fig. 1.3   Definitions of the start, middle, and end of a cycle.

The strain energy density increment per cycle is defined as the difference between the strain energy densities at the end and start of the cycle, i.e.,

$$\Delta\left(\frac{dW}{dV}\right) = \left(\frac{dW}{dV}\right)_{t_f} - \left(\frac{dW}{dV}\right)_{t_i} \tag{1.23}$$

As illustrated in Fig. 1.4, the strain energy density increment per cycle, $\Delta(dW/dV)$, can be related to the strain energy density factor increment, ΔS, by

$$\Delta\left(\frac{dW}{dV}\right) = \frac{\Delta S}{\Delta r} \tag{1.24}$$

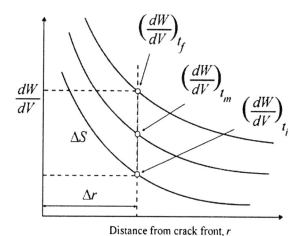

Fig. 1.4 Relationship between the strain energy density and the strain energy density factor.

Substituting from this equation into Eq. (1.21) results in the crack propagation rate

$$\frac{\Delta r}{\Delta N} = \frac{\Delta S}{\left(\dfrac{dW}{dV}\right)_c - \left(\dfrac{dW}{dV}\right)_{avg}} \tag{1.25}$$

in which ΔS and $(dW/dV)_{avg}$ are computed by using the stress and strain fields at a distance of Δr from the crack front. The critical value of the strain energy density, $(dW/dV)_c$, can be determined through an inverse approach by utilizing crack growth measurements in conjunction with finite element modeling and simulation, as discussed by Guven et al. (2001). The numerical value of the critical strain energy density is extracted as $(dW/dV)_{cr} = 571.33\,\text{MPa}$.

In conjunction with the finite element modeling and simulation, the number of thermal cycles for the specified number of crack growth increments can be obtained from Eq. (1.25) in the form

$$N = \sum_{i=1}^{n} \left\{ \frac{\left(\dfrac{dW}{dV}\right)_c - \left(\dfrac{dW}{dV}\right)_{avg}}{\Delta S} \Delta r \right\}_i \qquad (1.26)$$

in which i denotes the crack growth increment of Δr_i. At a certain stage during crack propagation, the slow crack growth due to cyclic loading ends and the onset of rapid crack growth begins when the last ligament of the material is reached. Sih and Moyer (1983) established that this occurs when the critical length parameter, r_c, is reached. This parameter is characteristic of the separation of the last ligament of material, triggering rapid crack growth. It is related to the critical value of the strain energy by

$$\left(\frac{dW}{dV}\right)_c = \frac{S_c}{r_c} \qquad (1.27)$$

in which S_c is the critical value of the strain energy density factor. The extent of the last ligament length prior to the onset of rapid crack growth requires extensive experimental investigation. Due to a lack of critical length data, the onset of rapid crack propagation can be assumed to commence when the last ligament length reaches 10% or 5% of the solder ball diameter.

1.4   References

Akay, H. U., Paydar, N. H., and Bilgic, A., 1997, "Fatigue Life Predictions for Thermally Loaded Solder Joints Using a Volume-Weighted Averaging Technique," *J. Electron. Packag.*, Vol. 119, pp. 228-235.

Anand, L., 1982, "Constitutive Equations for the Rate-Dependent Deformation of Metals at Elevated Temperatures," ASME *J. Mater. Technol.*, Vol. 104, pp. 12-17.

Anderson, T., Barut, A., Guven, I., and Madenci, E., 2000, "Revisit of Life Prediction Models for Solder Joints," *Proceedings, 50th Electronic Components and Technology Conference*, IEEE, New York, pp. 1064-1069.

Darveaux, R., 1996, "How to Use Finite Element Analysis to Predict Solder Joint Fatigue Life," *Proceedings, 6th International Congress on Experimental Mechanics*, Elsevier, New York, pp. 41-48.

Darveaux, R., 1997, "Solder Joint Fatigue Life Model," *Design and Reliability of Solder and Solder Interconnections, Proceedings of the TMS*, The Minerals, Metals, and Materials Society, Warrendale, pp. 213-218.

Darveaux, R., 2000, "Effect of Simulation Methodology on Solder Joint Crack Growth Correlation," *Proceedings, 50th Electronic Components and Technology Conference*, IEEE, New York, pp. 1048-1058.

Darveaux, R., Banerji, K., Mawer, A., and Dody, G., 1995, "Reliability of Plastic Ball Grid Array Assembly," *Ball Grid Array Technology* (J. Lau, Editor), McGraw-Hill, Inc., New York, pp. 379-442.

Darveaux, R. and Syed, A., 2000, "Reliability of Area Array Solder Joints in Bending," *SMTA International Proceedings 2000*, Surface Mount Technology Association, Edina, MN, pp. 313-324.

Gustafsson, G., Guven, I., Kradinov, V., and Madenci, E., 2000, "Finite Element Modeling of BGA Packages for Life Prediction," *Proceedings, 50th Electronic Components and Technology Conference*, IEEE, New York, pp. 1059-1063.

Guven, I., Kradinov, V., and Madenci, E., 2001, "Strain Energy Density Criterion for Reliability Life Prediction of Solder Joints in Electronic Packaging," *ASME J. Electron. Packag.* (submitted).

Hall, P. M. and Sherry, W. M., 1986, "Materials, Structures and Mechanics of Solder Joints for Surface-Mount Microelectronics Technology," *Proceedings, Conference of Interconnection Technology in Electronics*, Welding Society of Germany, Düsseldorf, Germany, pp. 47-61.

Iannuzzelli, R. J., Pitarresi, J. M., and Prakash, V., 1996, "Solder Joint Reliability Prediction by the Integrated Matrix Creep Method," *J. Electron. Packag.*, Vol. 118, pp. 55-61.

Knecht, S. and Fox, L. R., 1990, "Constitutive Relation and Creep-Fatigue Life Model for Eutectic Tin-Lead Solder," *IEEE Trans. Compon. Packag. Manuf. Technol.*, Vol. 13, pp. 424-433.

Lam, S. T., Arieli, and Mukherjee, A. K., 1979, "Superplastic Behavior of Pb-Sn Eutectic Alloy," *Mater. Sci. Eng.*, Vol. 40, pp. 73-79.

Lau, J., Chang, C., and Lee, S. W. R., 2000, "Solder Joint Crack Propagation Analysis of Wafer-Level Chip Scale Package on Printed Circuit Board Assemblies," *Proceedings, 50th Electronic Components and Technology Conference*, IEEE, New York, pp. 1360-1368.

Lee, W. W., Nguyen, L. T., and Selvaduray, G. S., 2000, "Solder Joint Fatigue Models: Review and Applicability to Chip Scale Packages," *Microelectron. Reliab.*, Vol. 40, pp. 231-244.

Liu, Y., Irving, S., Tumulak, M. and Cabahug, E. A., 2002, "Assembly Process Induced Stress Analysis for New FLMP Package by 3D FEA," *Proceedings, 52th Electronic Components and Technology Conference*, IEEE, New York, paper s14p4 (on CD).

Mukherjee, A. K., Bird, J. E., and Dorn, J. E., 1969, "Experimental Correlations for High-Temperature Creep," *Trans. Am. Soc. Metals*, Vol. 62, pp. 155-179.

Pan, T., 1994, "Critical Accumulated Strain Energy (CASE) Failure Criterion for Thermal Cycling Fatigue of Solder Joints," *J. Electron. Packag.*, Vol. 116, pp. 163-170.

Sih, G. C., 1973, "Some Basic Problems in Fracture Mechanics and New Concepts," *Eng. Fract. Mech.*, Vol. 5, pp. 365-377.

Sih, G. C., 1991, *Mechanics of Fracture Initiation and Propagation*, Kluwer Academic Publishers, New York.

Sih, G. C. and MacDonald, B., 1974, "Fracture Mechanics Applied to Engineering Problems—Strain Energy Density Criterion," *Eng. Fract. Mech.*, Vol. 6, pp. 361-386.

Sih, G. C. and Moyer, E. T., Jr. 1983, "Path Dependent Nature of Fatigue Crack Growth," *Eng. Fract. Mech.*, Vol. 17, pp. 269-280.

Syed, A., 1997, "Factors Affecting Creep-Fatigue Interaction in Eutectic Sn/Pb Solder Joints," *Advances in Electronic Packaging*, ASME, New York, EEP-Vol. 19-2, pp. 1535-1532.

Syed, A., 2001, "Predicting Solder Joint Reliability for Thermal, Power, & Bend Cycle Within 25% Accuracy," *Proceedings, 51st Electronic Components and Technology Conference*, IEEE, New York, pp. 255-263.

Chapter 2

THERMOMECHANICAL FATIGUE LIFE PREDICTION ANALYSIS

The life prediction of solder joints under thermal cyclic loading requires a two-stage analysis: (1) a three-dimensional finite element analysis to compute the stress, strain, and strain energy field, and (2) a life prediction analysis. As shown in Fig. 2.1, a typical thermal cycle is described by its starting, maximum, and minimum temperatures, as well as the durations of time for ramp-up, ramp-down, and dwell at maximum and minimum temperatures. The starting temperature is the reference (stress-free) temperature of the thermal cycle. Four thermal cycles in the simulation are usually sufficient to ensure the stability of the hysteresis loop.

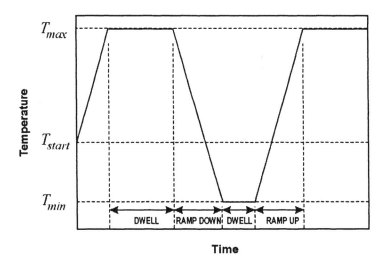

Fig. 2.1   Description of a thermal cycle (ramp-up, -down, dwell).

2.1 Approach

Various finite element modeling techniques exist for predicting the fatigue life of a solder joint in a ball grid array (BGA)-type electronic package [see Gustafsson et al. (2000) for a comparison of the techniques]. These common modeling techniques can be classified into five categories.

Method 1: Nonlinear Slice Model - This commonly accepted modeling technique, applicable to packages having octant symmetry, utilizes only a diagonal slice of the assembly in order to reduce computation time. The slice passes through the thickness of the assembly, capturing all major components and a full set of solder joints. The model imposes symmetry boundary conditions on the slice plane coinciding with the true symmetry plane. The other slice plane is neither a free surface nor a true symmetry plane. This modeling difficulty is resolved by requiring that this slice plane remain parallel to the symmetry plane. However, this requirement implies that the package is infinitely long in the direction perpendicular to the slice plane. This leads to an underestimation of the warpage of the package/board during the temperature cycling or an overestimation of the shear loading on the solder joints. Thus, it results in underprediction of the thermal cycle life. The applied degree-of-freedom constraints on the slice model are shown in Fig. 2.2. The temperature-dependent material properties and viscoplastic solder behavior are included in the three-dimensional nonlinear analysis under appropriate thermal cycles.

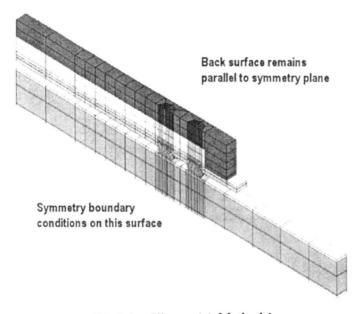

Back surface remains
parallel to symmetry plane

Symmetry boundary
conditions on this surface

Fig. 2.2   Slice model–Method 1.

Method 2: Nonlinear Global Model with Linear Super Elements - This approach avoids the assumptions associated with the boundary conditions of the slice model. As shown in Fig. 2.3, the package and board are modeled as

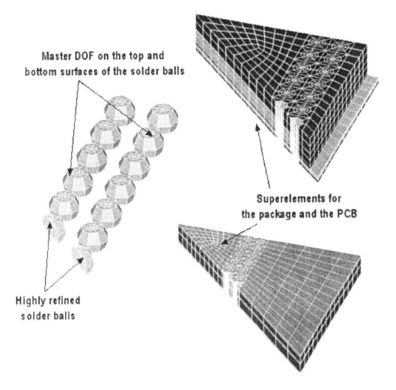

Master DOF on the top and
bottom surfaces of the solder balls

Superelements for
the package and the PCB

Highly refined
solder balls

Fig. 2.3   Global model with superelements–Method 2.

two super-elements, and all the solder balls as three-dimensional finite elements. Except for the critical joint(s), the solder balls are modeled with a coarse mesh. The critical solder joint identified *a priori* is modeled with a highly refined mesh. The super-elements include only the linear material properties, and all the solder joints exhibit nonlinear material behavior. The global model, in which the package and the board are represented by the super elements, is subjected to the appropriate thermal cycles.

Method 3: Linear Global Model with a Nonlinear Submodel - An alternative to the preceding approach is a linear model of the substrate, board, and all the solder balls using three-dimensional finite elements in order to identify the critical solder joint for a subsequent nonlinear submodeling. Refinements of the mesh in the global model and in the submodel are shown in Fig. 2.4. The global model includes only the linear material properties, whereas the submodel includes nonlinear material behavior. The linear global model is subjected to a one-degree temperature change, providing displacement fields on a per-degree basis. The scaled displacement fields in accordance with the thermal cycling become the boundary conditions for the submodel of the critical joint. This permits the simulation of any thermal

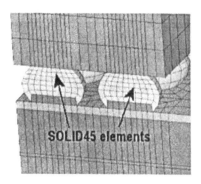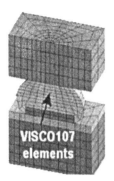

Fig. 2.4   Linear global model with nonlinear submodel–Method 3.

cycle using only one set of global model results. This approach was suggested by Riebling (1996).

Method 4: Nonlinear Global Model with a Nonlinear Submodel - This approach is similar to the preceding approach except for the linear material properties in the global model. The nonlinear global model, with a very coarse mesh for the substrate, board, and solder balls, provides the critical joint for the subsequent nonlinear submodeling of the critical solder joint.

Mesh refinement in the submodel is the same as that in the previous model. As in the preceding approach, the displacements become the boundary conditions for the nonlinear submodel of the critical joint in accordance with the thermal cycling. The global and submodel include temperature-dependent properties and the viscoplastic material behavior.

Method 5: Nonlinear Global Model - The global model employs a relatively coarse mesh for all the components of a package except for the critical solder joint(s). However, the mesh generation aspect of this approach becomes time consuming because the fine mesh associated with the critical solder joint(s) must match the coarse mesh of the remaining solder joints. The model includes temperature-dependent material properties and the viscoplastic behavior of the solder joints. If the package consists of a large number of solder joints, it is not feasible to model all the solder joint(s) with a refined mesh.

According to a comparison of life predictions utilizing the criterion by Darveaux (1997), Gustafsson et al. (2000) found that Method 2 is not acceptable. The prediction by Method 1 is conservative in comparison to the others. Considering the modeling assumptions, the results from Methods 3, 4, and 5 indicate that Method 4 is likely to have more fidelity than the others.

The submodeling stage of the analysis involves the application of the corresponding displacement boundary conditions to the surfaces that define the interface between the global model and the submodel. The displacement boundary conditions are determined from the solution of the global analysis through the use of the cut-boundary interpolation method. Once the interpolated displacement field on the submodel is known, it is used in the simulation of the complete thermal cycle until a stable stress-strain hysteresis loop is achieved.

In the nonlinear analysis, viscoplastic solid finite elements (VISCO107) with Anand's nonlinear material response (built into the ANSYS® software) are used. Whenever these elements are used, the large-deformation-effects option is enabled (NLGEOM,ON). The constants required by the built-in ANSYS® material model are shown in Table 1.3. The package components, except for the solder, are modeled using SOLID45 elements appropriate for linear material behavior.

2.2 Analysis Steps

This section describes the analysis steps for creating a *nonlinear global model with a nonlinear submodel*. The user must have detailed geometric information about the actual package. A micro-photograph of the sectioned package, similar to the one shown in Fig. 2.5, would prove extremely useful as it would provide valuable information on the local geometry of the solder joints. Geometric symmetry conditions must be identified. The user must keep in mind that if there is symmetry, a significant amount of time will be saved in the model generation and in the solution stages of the analysis. The necessary geometrical information includes the pitch, copper pad radii and thicknesses, standoff height, and all dimensions (length, width, and thickness) of every component in the package.

Fig. 2.5   Section photo of a solder ball from TI's Micro-Star.

As in every FEM-based analysis, the analysis starts with the model generation, followed by the simulation of the thermal cycles, transition from the
global model to the submodel, and fatigue life calculation. The basic steps
of the analysis are

- Global model generation
- Apply symmetry conditions
- Apply displacement constraints on one node in all directions
- Apply thermal load as a function of time (simulation of the thermal
  cycles)
- Solve
- Identify most critical solder joint
- Save global model results
- Submodel generation (most critical solder joint)
- Apply thermal load and cut-boundary displacement conditions as a
  function of time (from global model results)
- Solve
- Select elements at possible crack initiation sites
- Find volume-weighted plastic work for selected elements
- Calculate fatigue life prediction

ReliANS, a group of ANSYS® macros, was designed/developed to carry out
these analysis steps in a seamless fashion within the ANSYS®-GUI environment. These macros, not available in the standard ANSYS® program, must
be installed by following the instructions provided in Appendix A. A
detailed explanation of each macro is given in Chap. 4.

2.3 Case Study: BGA-Type Package

2.3.1 Data Needed Before Beginning ANSYS® Session

A demonstration of the use of ReliANS commands is given by considering a
BGA-type package with 96 solder balls arranged in a peripheral pattern. As
shown in Fig. 2.6, there are 12 and 14 balls in each direction (x and z) along
each side of the inner and outer rings, respectively. All the dimensions are
given in millimeters unless otherwise specified. The pitch is 0.5 mm. Die
and substrate sizes are given as $5.2 \times 5.2 \times 0.3$ mm^3 and $8.0 \times 8.0 \times 0.075$
mm^3, respectively. A cross-sectional view is illustrated in Fig. 2.7, with
labels identifying the component. Figure 2.8 shows the thickness information that is needed in the modeling phase. The x and z dimensions of the
PCB are not given since they are arbitrary. In such a case, the common practice is to assume a PCB size of at least twice the size of the substrate. In this
case, the PCB size is taken as 20 mm × 20 mm. Also, a cross-sectional

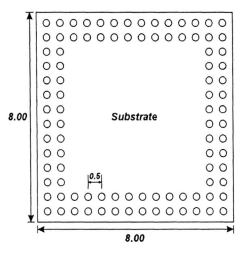

Fig. 2.6   Solder joint patterns for the BGA with 96 balls.

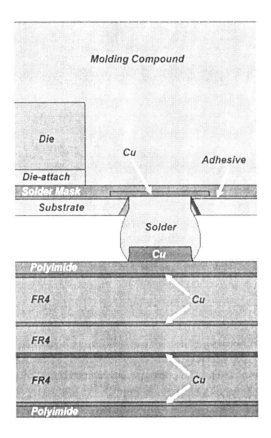

Fig. 2.7   Cross-sectional view of the package on the PCB.

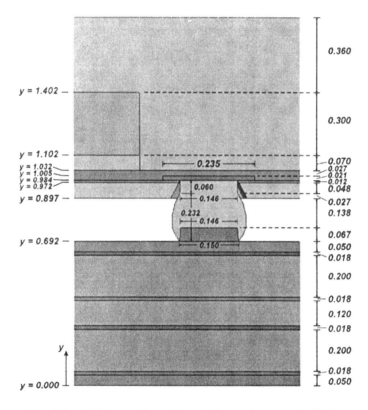

Fig. 2.8   Thickness dimensions of the package on the PCB.

micro-photograph of a typical solder ball (see Fig. 2.5) is used in extracting the required information for the curved surface of the solder ball. Once all the geometrical and material information is ready, the user can start the modeling phase with the global model. Thermal cyclic loading varies between -40 and 100°C, with 4-min ramp-up, 6-min ramp-down, and dwell times of 26 min at maximum temperature and 24 min at minimum temperature. The reference (stress-free) temperature is taken to be 25°C. Simulation of four cycles is sufficient to ensure the stability of the hysteresis loop.

2.3.2 Global Model Generation

The geometrical parameters defining the solder joint are given in section 2.3.1. Starting with data set 1, specify these parameters as given in Fig. 2.9. Data sets 2 and 3 are illustrated in Figs. 2.10 and 2.11, respectively. If the user prefers to use ANSYS® for the command line input, then the location of the ReliANS macro directory and the element type should be declared in the

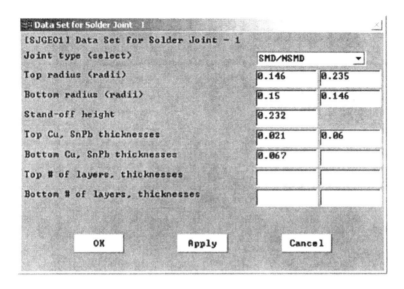

Fig. 2.9 Data set 1 for solder joint(s) (SJGEO1 command).

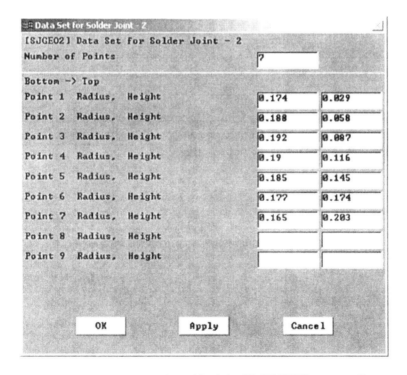

Fig. 2.10 Data set 2 for solder joint(2) (SJGEO2 command).

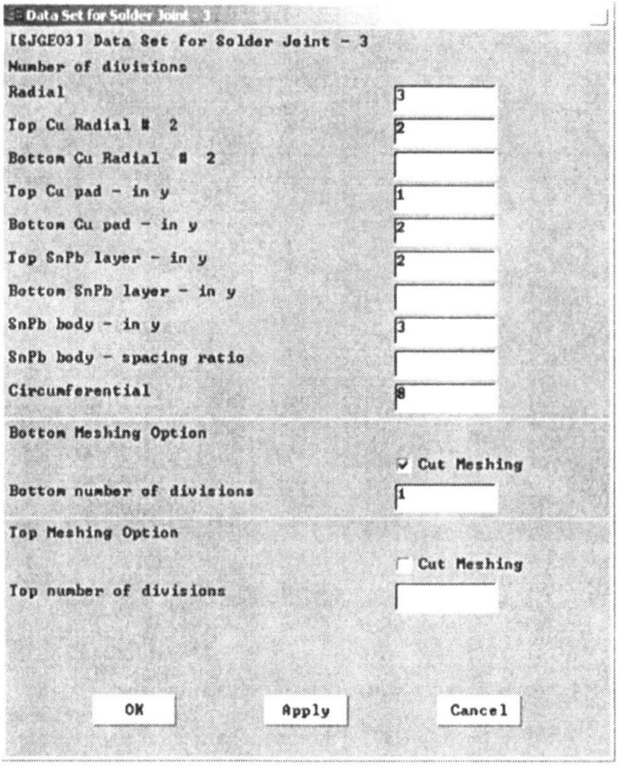

Fig. 2.11   Data set 3 for solder joint(s) (SJGEO3 command).

input window. These steps are done automatically when the user enters the *"Packaging"* menu under the *"Preprocessor"* in the ANSYS®-GUI. The command line input for these steps is given as

```
/PSEARCH,C:\RELIANS
/PREP7
ET,1,45
ET,2,42
```

The command "/PSEARCH" must be issued for each new ANSYS® session.

The command line input[1] for the definition of the geometrical parameters of the solder balls is

```
SJGEO1,2,0.146,0.235,0.15,0.146,0.232,0.021,0.06,0.067,,,,,
SJGEO2,7,0.174,0.029,0.188,0.058,0.192,0.087,0.19,0.116,0.185,
     0.145,0.177,0.174,0.165,0.203
SJGEO3,3,2,,1,2,2,,,3,,8,1,1,,
```

---

[1]An indented line is a continuation of the preceding command line.

Once the geometrical parameters for the solder joint have been specified, the user is ready to create the solder joint array. This is achieved by using SJFULL, as shown in Fig. 2.12. The command line input for this action is

SJFULL,2.75,0.692,0.25,0.5,2,7,1,2,0

Different views (left oblique, top, and front) of the mesh created by execution of this command are shown in Fig. 2.13. Note that the starting x and z positions are specified as 2.75 mm and 0.25 mm (5.5 times the pitch and half the pitch), respectively. The user has to keep in mind that the **global origin** in ANSYS® corresponds to the **center** of the package.

After the mesh for the solder joints (and copper pads) has been created, the mesh related to the PCB is considered. For this, the user must first provide the geometrical and material information for the individual layers in the PCB. The first step involves the layer thickness information, which is specified using FR4GEO1. The data entered for this command are shown in Fig. 2.14. The numbers of divisions in the y-direction are specified using FR4GEO2. The data entered for this command are shown in Fig. 2.15. The material numbers for each layer are specified using FR4GEO3. The data entered for this command are shown in Fig. 2.16. Note that four different materials are specified in FR4GEO3 (Fig. 2.16), i.e., materials 2, 3, 4, and 5. Materials 2 and 3 correspond to the same material, copper. This is done intentionally in order to demonstrate the usage of this command and its capability of handling an arbitrary number of materials in an arbitrary sequence.

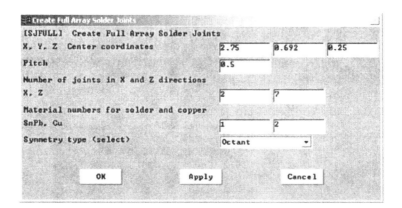

Fig. 2.12 Data used to create the array of solder joints (SJFULL command).

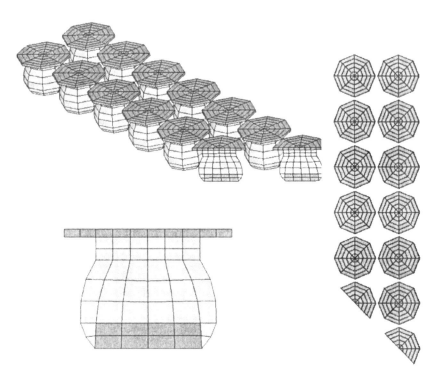

Fig. 2.13   Mesh created by the SJFULL command: front view at lower left; top view at right (not to scale).

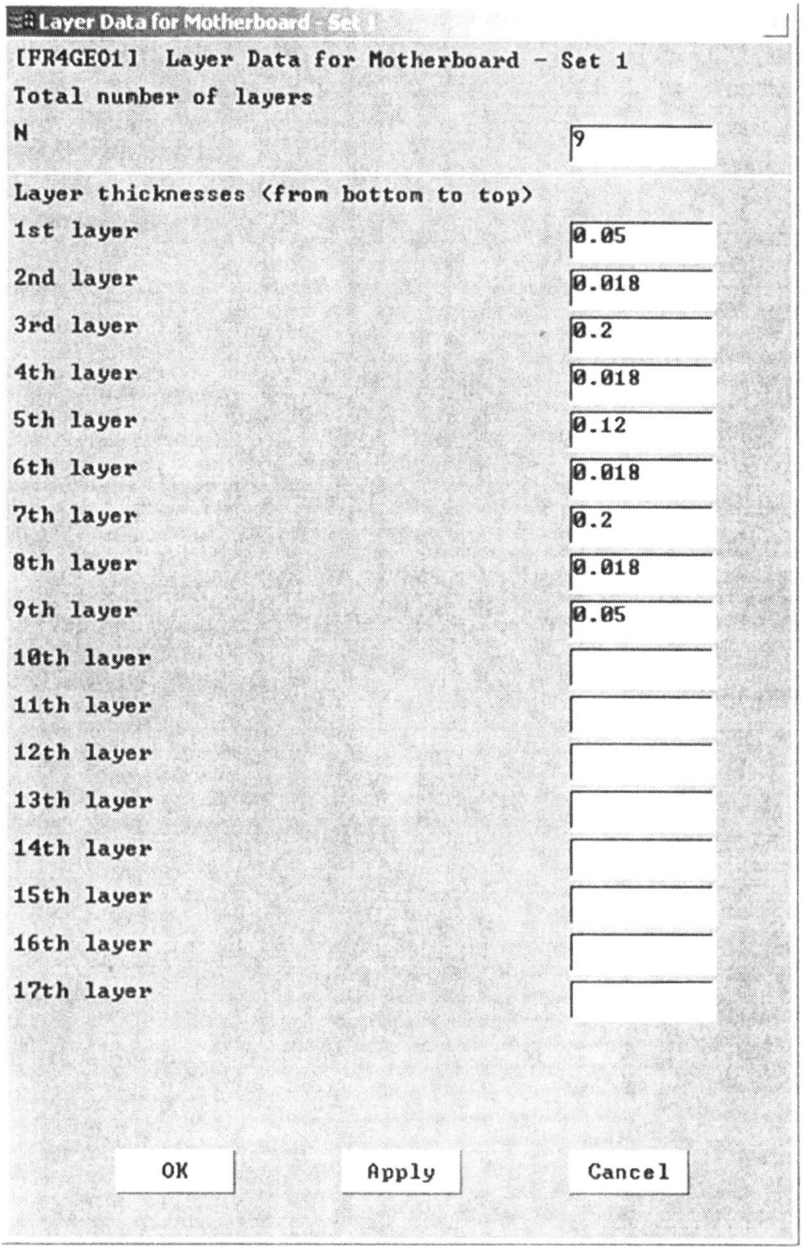

Layer Data for Motherboard - Set 1	
[FR4GEO1] Layer Data for Motherboard - Set 1	
Total number of layers	
N	9
Layer thicknesses (from bottom to top)	
1st layer	0.05
2nd layer	0.018
3rd layer	0.2
4th layer	0.018
5th layer	0.12
6th layer	0.018
7th layer	0.2
8th layer	0.018
9th layer	0.05
10th layer	
11th layer	
12th layer	
13th layer	
14th layer	
15th layer	
16th layer	
17th layer	

OK	Apply	Cancel

Fig. 2.14 Data set 1 for motherboard layers (FR4GEO1 command).

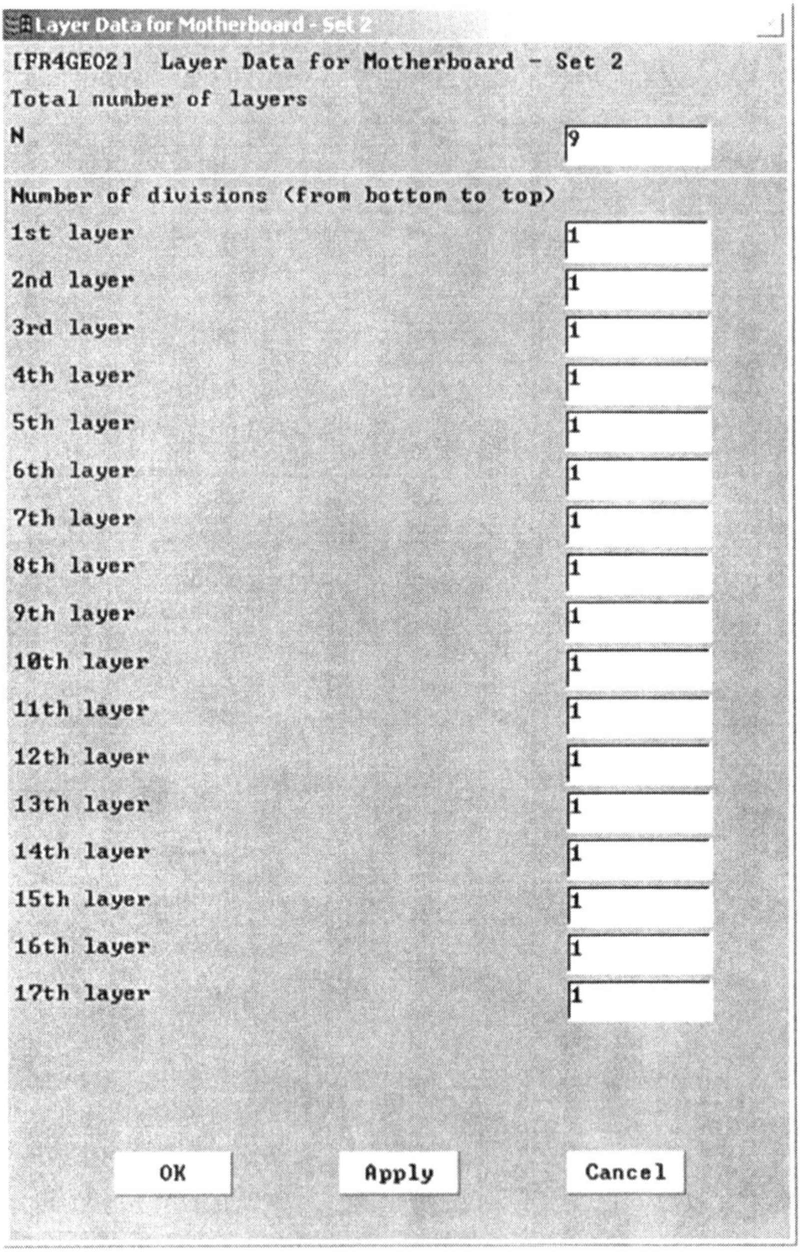

Fig. 2.15   Data set 2 for motherboard layers (FR4GEO2 command).

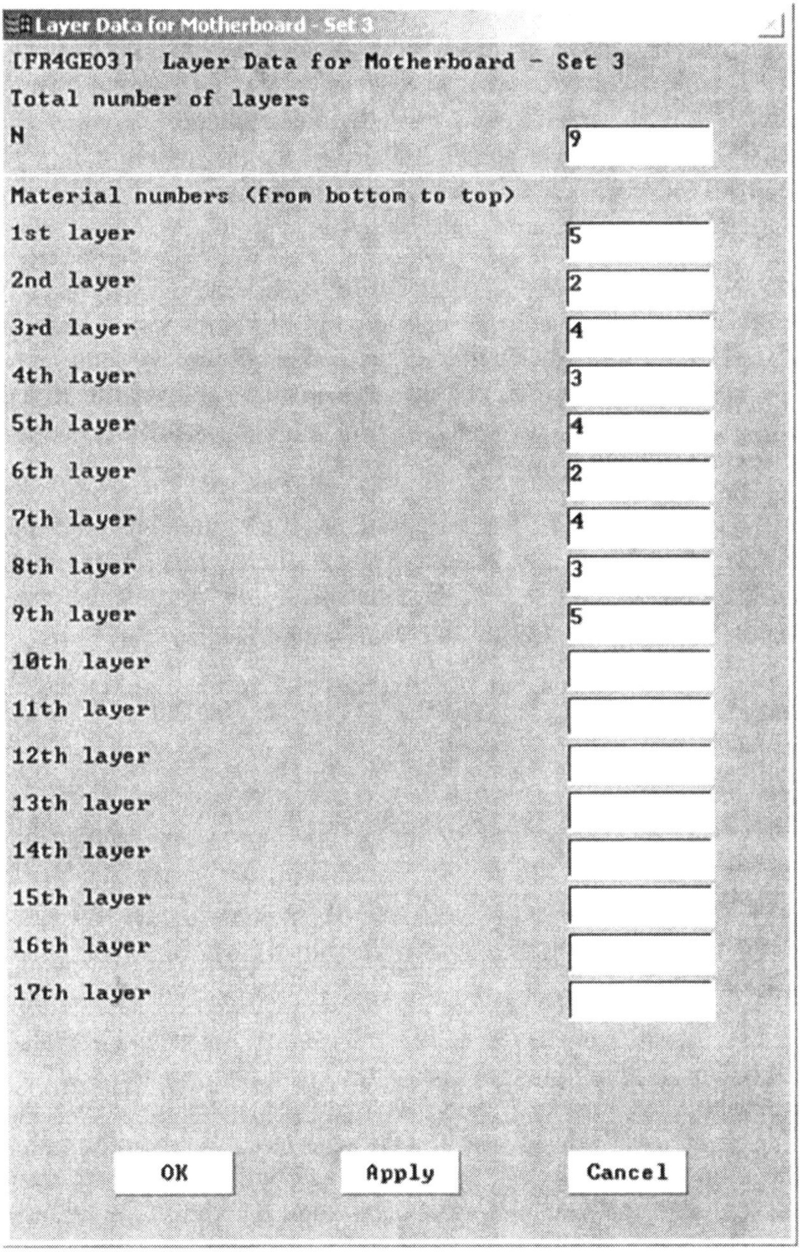

Fig. 2.16 Data set 3 for motherboard layers (FR4GEO3 command).

Once the layer geometry and material information have been specified, the user can create a PCB mesh at any location in any of the available patterns and shapes. First, the PCB mesh underneath the solder joints is created using BLCK7 with the parameters specified as in Fig. 2.17. After issuing this command, the user should see a mesh that looks similar to the one given in Fig. 2.18 (top and bottom oblique). Note that the solder joint radius and the number of radial divisions are consistent with the existing solder joint mesh pattern (bottom of the solder joint).

Next, the area outside the solder joint is created using BLCK9 (Fig. 2.19). The status of the mesh after the execution of BLCK9 is shown in Fig. 2.20. In Fig. 2.19, note that the number of divisions in the z-direction is specified as 14. The reason for this is to create a mesh that will match the PCB mesh underneath the solder joint area. If the user does not ensure this consistency, the results will be **wrong**.

Finally, the triangular PCB mesh inside the solder joint area is created by using BLCK10 (Fig. 2.21). The mesh status as it appears after execution of BLCK10 is shown in Fig. 2.22. The command line input for the three data sets and the creation of three different PCB regions is given as

```
FR4GEO1,9,0.05,0.018,0.2,0.018,0.12,0.018,0.2,0.018,0.05
FR4GEO2,9,1,1,1,1,1,1,1,1,1
FR4GEO3,9,5,2,4,3,4,2,4,3,5

BLCK7,2.75,0.0,0.25,0.5,2,7,0.15,,2,,,8,0
BLCK9,3.5,0.0,0.0,6.5,10,45,5,14
BLCK10,0,0,0,2.5,10
```

Figure 2.23 shows the mesh status around the solder joints after the mesh generations for the solder joints (with copper pads) and PCB have been completed.

The next step is to create the adhesive layer that surrounds the solder joint neck region. Industry practice shows that this adhesive layer does not give much support to the solder once thermal cycling starts. In order to account for this behavior, the inside walls of the adhesive are specified to be slanted, with a 60° slope. Figure 2.24 shows the dialog box for this command (BLCK2) with the specified values, and Figs. 2.25 and 2.26 (oblique and front view, respectively) show the mesh created upon execution. Next, the mesh outside the solder joint area is created by using BLCK4 as given in Fig. 2.27. The mesh created as a result of this action is shown in Fig. 2.28. Similarly, the portion of the adhesive inside the solder joint area (triangular) is created using BLCK5 as given in Fig. 2.29, resulting in the mesh shown in Fig. 2.30. The command line input for the generation of adhesive mesh is given as

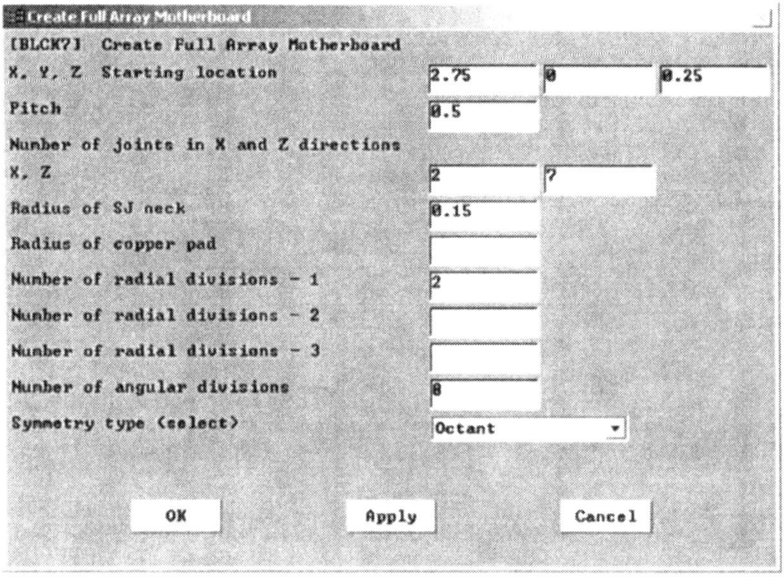

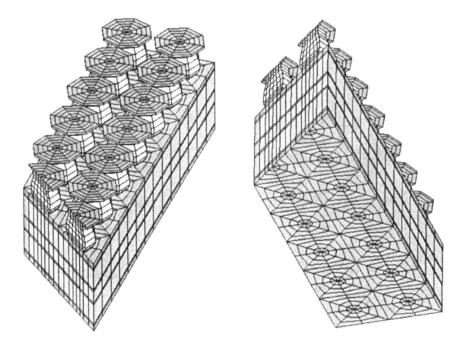

Create Full Array Motherboard

[BLCK7] Create Full Array Motherboard

X, Y, Z Starting location	2.75	0	0.25
Pitch	0.5		

Number of joints in X and Z directions

X, Z	2	7

Radius of SJ neck	0.15
Radius of copper pad	
Number of radial divisions - 1	2
Number of radial divisions - 2	
Number of radial divisions - 3	
Number of angular divisions	8
Symmetry type (select)	Octant ▼

OK   Apply   Cancel

Fig. 2.17   Data used to create the array PCB mesh with a solder joint mesh
pattern (BLCK7 command).

Fig. 2.18   Top (left) and bottom (right) oblique views of the mesh created by the
BLCK7 command.

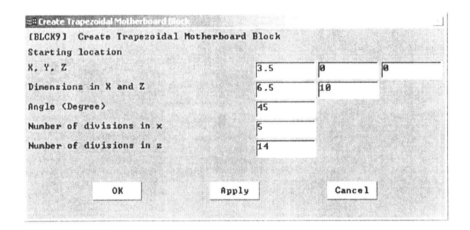

Fig. 2.19   Data used to create the trapezoidal PCB block outside
the solder joint area (BLCK9 command).

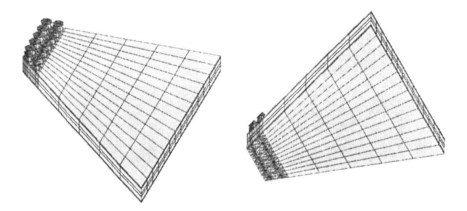

Fig. 2.20   Top (left) and bottom (right) oblique view of the mesh after the
BLCK9 command is executed.

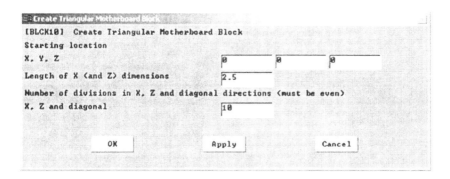

Fig. 2.21 Data used to create the triangular PCB mesh inside the solder joint area (BLCK10 command).

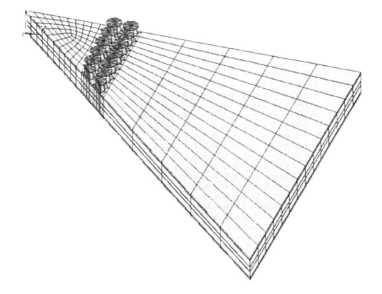

Fig. 2.22 Oblique view of the mesh after the BLCK10 command is executed.

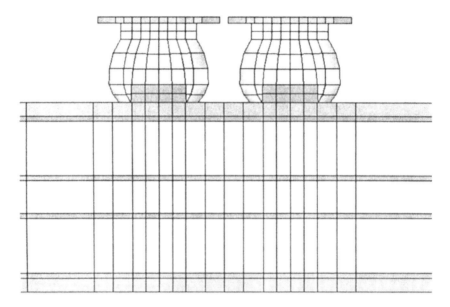

Fig. 2.23   Front view of the mesh around the solder joints after the joints
           and PCB mesh are created.

Fig. 2.24   Data used to create the adhesive mesh around the solder
           area (BLCK2 command).

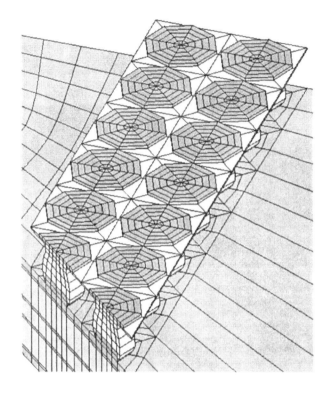

Fig. 2.25   Oblique view of the mesh after the BLCK2 command is executed.

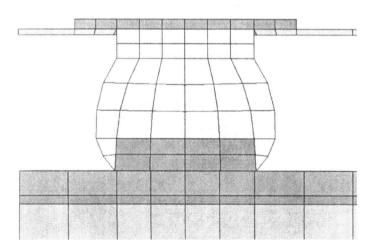

Fig. 2.26   Front view of the mesh after the BLCK2 command is executed.

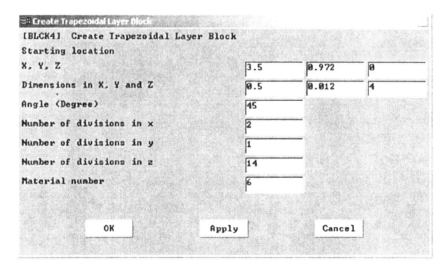

Fig. 2.27   Data used to create the adhesive mesh outside the solder joint area
(BLCK4 command).

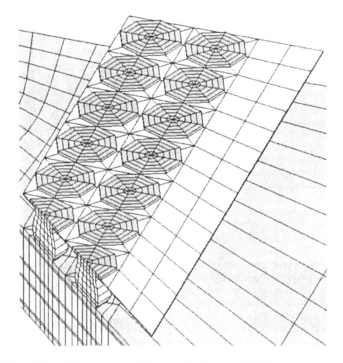

Fig. 2.28   Oblique view of the mesh after the BLCK4 command
is executed.

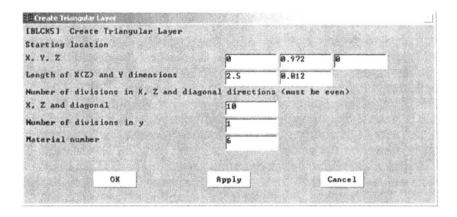

Fig. 2.29   Data used to create the adhesive mesh inside the solder joint area (BLCK5 command).

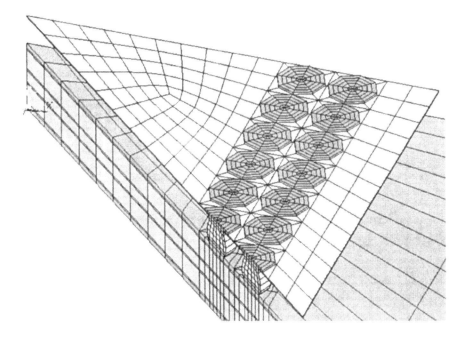

Fig. 2.30   Oblique view of the mesh after the BLCK5 command is executed.

```
BLCK2,2.75,0.972,0.25,0.5,2,7,0.146,0.235,0.012,60,2,1,1,8,6,0
BLCK4,3.5,0.972,0.0,0.5,0.012,4,45,2,1,14,6
BLCK5,0.0,0.972,0.0,2.5,0.012,10,1,6
```

The substrate mesh is generated next. The procedure is exactly the same as that for the die-attach and, therefore, the details are not repeated here. However, for the sake of completeness, the dialog boxes for the commands that are executed and the resulting meshes at each stage are given in Figs. 2.31-2.36. Note that the radius of the solder area is in the form of a parameter. This is because the substrate, with slanted inner walls, is located underneath the adhesive layer and because the difference (increase) in the radius value that is caused by the adhesive thickness must be taken into account. This difference is calculated by the operation shown in Fig. 2.31, and the newly calculated radius (param1) for the solder joint area in the substrate is used in BLCK2. The command line input for the generation of the substrate mesh is given as

```
PARAM1=0.146+0.012/TAN(ATAN(1)*4/3)
BLCK2,2.75,0.897,0.25,0.5,2,7,PARAM1,0.235,0.075,60,2,1,2,8,
    7,0
BLCK4,3.5,0.897,0.0,0.5,0.075,4,45,2,2,14,7
BLCK5,0.0,0.897,0.0,2.5,0.075,10,2,7
```

The solder mask mesh is generated next. This operation is done in two steps because part of the solder mask surrounds the copper pads and the other part completely covers the package cross-section. In other words, two layers of solder mask are created. The generation of the mesh is very similar to that described so far, and the details are not repeated here. Figures 2.37-2.45 give the dialog boxes for the commands used and corresponding resulting mesh status. The command line input for the generation of the first layer of the solder mask is given as

```
BLCK2,2.75,0.984,0.25,0.5,2,7,0.235,,0.021,90,3,1,1,8,8,0
BLCK4,3.5,0.984,0.0,0.5,0.021,4,45,2,1,14,8
BLCK5,0.0,0.984,0.0,2.5,0.021,10,1,8
```

The command line input for the generation of the second layer of the solder mask is given as

```
BLCK1,2.75,1.005,0.25,0.5,2,7,0.146,0.235,0.027,3,2,1,1,8,8,0
BLCK4,3.5,1.005,0.0,0.5,0.027,4,45,2,1,14,8
BLCK5,0.0,1.005,0.0,2.5,0.027,10,1,8
```

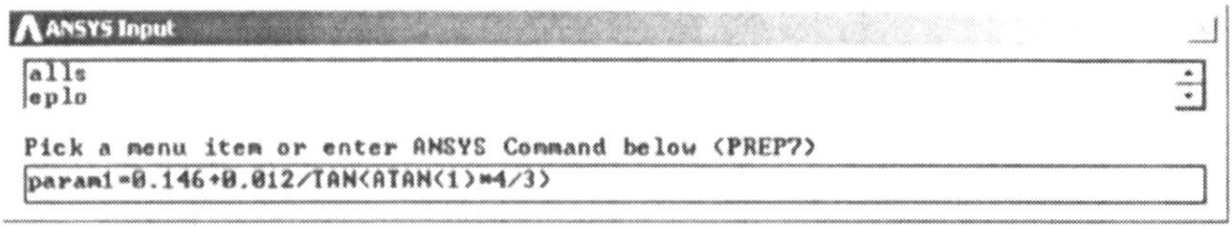

Fig. 2.31   Parameter definition for the radius at the slanted wall of the substrate.

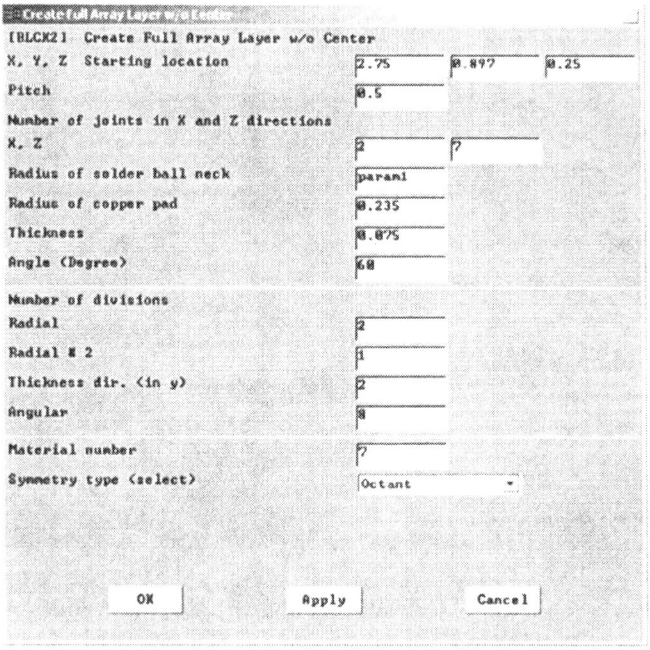

Create Full Array Layer w/o Center

[BLCK2] Create Full Array Layer w/o Center

X, Y, Z Starting location	2.75	0.897	0.25
Pitch	0.5		

Number of joints in X and Z directions

X, Z	2	7
Radius of solder ball neck	param1	
Radius of copper pad	0.235	
Thickness	0.075	
Angle (Degree)	60	

Number of divisions

Radial	2
Radial # 2	1
Thickness dir. (in y)	2
Angular	8
Material number	7
Symmetry type (select)	Octant ▾

OK   Apply   Cancel

Fig. 2.32   Data used to create the substrate mesh around the solder joint area (BLCK2 command).

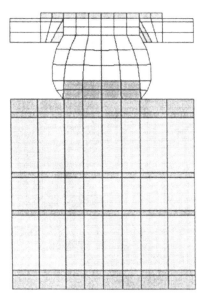

Fig. 2.33   Front view of the mesh after the BLCK2 command is executed.

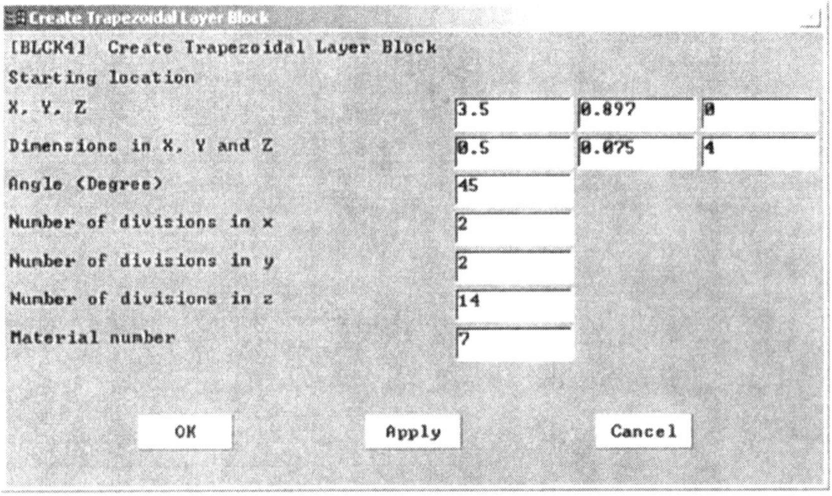

Fig. 2.34   Data used to create the substrate mesh outside the solder joint
              area (BLCK4 command).

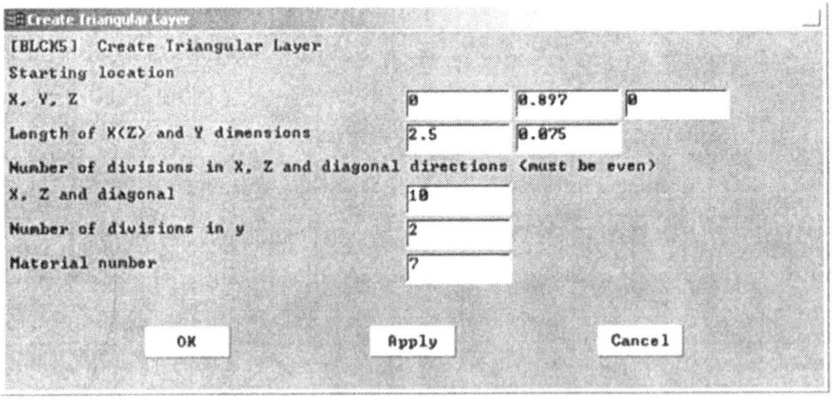

Fig. 2.35   Data used to create the substrate mesh inside the solder joint
              area (BLCK5 command).

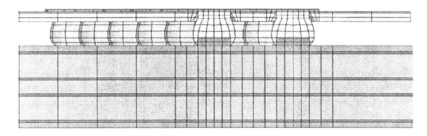

Fig. 2.36 Front view of the mesh after the substrate mesh is completed.

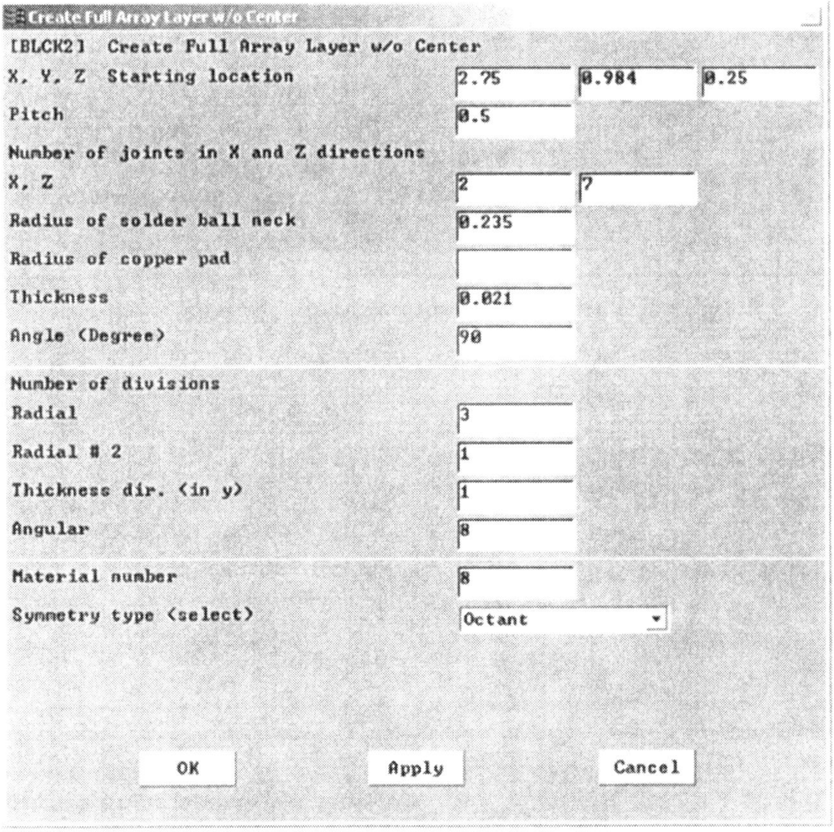

Create Full Array Layer w/o Center
[BLCK2] Create Full Array Layer w/o Center
X, Y, Z Starting location
2.75
0.984
0.25
Pitch
0.5
Number of joints in X and Z directions
X, Z
2
7
Radius of solder ball neck
0.235
Radius of copper pad
Thickness
0.021
Angle (Degree)
90
Number of divisions
Radial
3
Radial # 2
1
Thickness dir. (in y)
1
Angular
8
Material number
8
Symmetry type (select)
Octant
OK
Apply
Cancel

Fig. 2.37 Data used to create the solder mask mesh around the solder joints (BLCK 2 command).

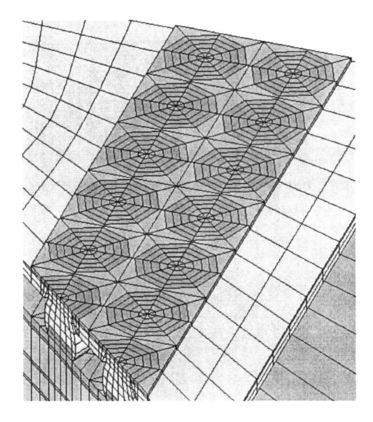

Fig. 2.38   Oblique view of the mesh after the BLCK2 command
            is executed.

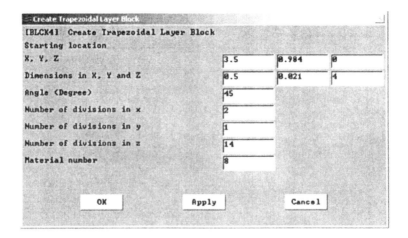

Fig. 2.39   Data used to create the solder mask mesh outside the solder
            joint area (BLCK4 command).

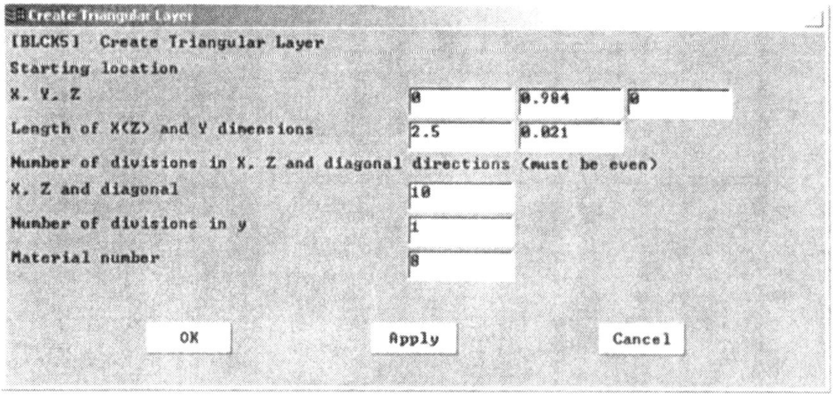

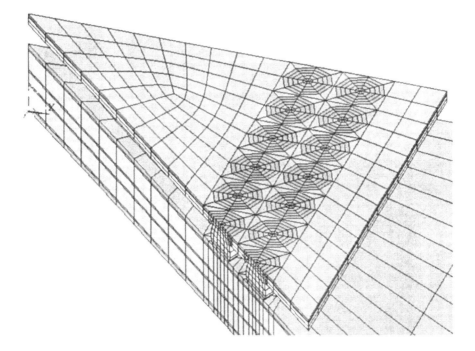

Create Triangular Layer

[BLCK5] Create Triangular Layer

Starting location

X, Y, Z   0   0.984   0

Length of X(Z) and Y dimensions   2.5   0.021

Number of divisions in X, Z and diagonal directions (must be even)

X, Z and diagonal   10

Number of divisions in y   1

Material number   8

OK   Apply   Cancel

Fig. 2.40 Data used to create the solder mask mesh inside the solder joint area (BLCK5 command).

Fig. 2.41 Oblique view of the mesh after the first layer of the solder mask mesh is completed.

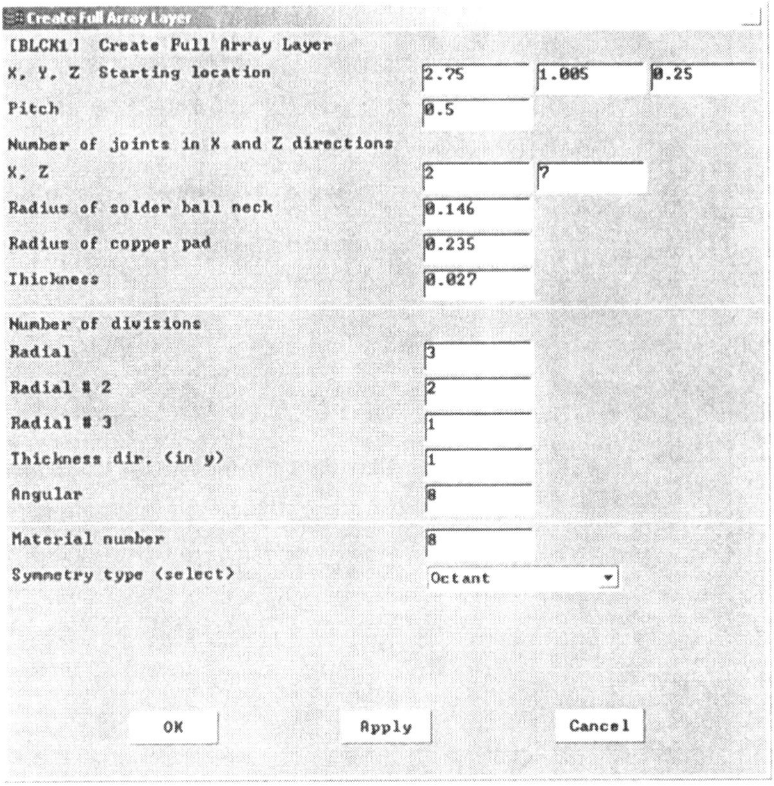

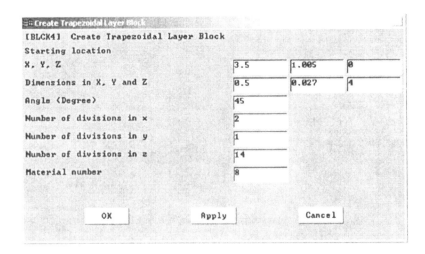

Fig. 2.42   Data used to create the solder mask mesh above the solder joint area
(BLCK1 command).

Fig. 2.43   Data used to create the solder mask mesh outside the solder joint area
(BLCK4 command).

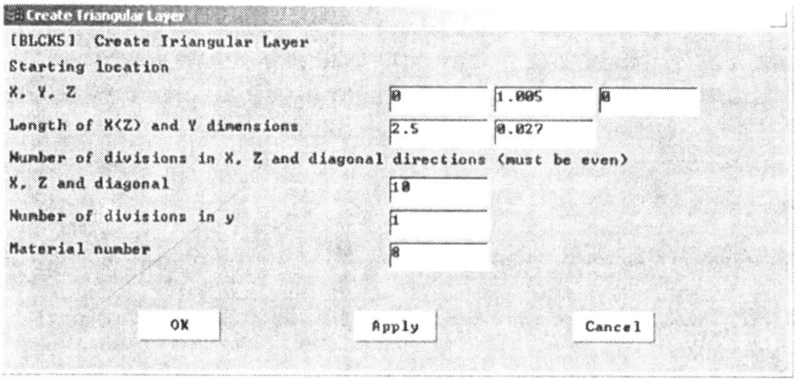

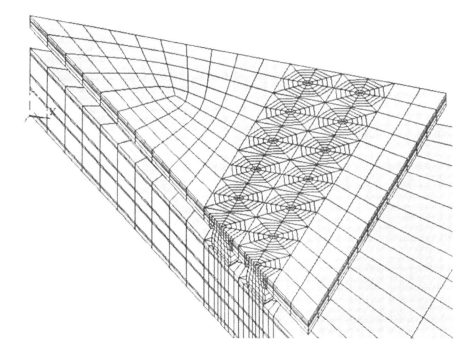

Create Triangular Layer

[BLCK5] Create Triangular Layer
Starting location
X, Y, Z   0   1.005   0
Length of X(Z) and Y dimensions   2.5   0.027
Number of divisions in X, Z and diagonal directions (must be even)
X, Z and diagonal   10
Number of divisions in y   1
Material number   8

            OK   Apply   Cancel

Fig. 2.44   Data used to create the solder mask mesh inside the solder joint area
              (BLCK5 command).

Fig. 2.45   Oblique view of the mesh after the second (last) layer of the solder
              mask mesh is completed.

Generation of the meshes for the die-attach and die is discussed next. The actual dimension for the die is given as 5.2 mm square. However, with the current mesh configuration, in order to be able to create a mesh with straight edges for the die, one must approximate a die size slightly different from the reality, i.e., 5 mm square in this case. Since the mesh generation is very similar to the ones described previously, the details are omitted here. Figures 2.46-2.49 give the dialog boxes with appropriate parameters and the resulting mesh configurations. The command line input for the generation of the die-attach is given as

```
BLCK5,0.0,1.032,0.0,2.5,0.07,10,1,9
```

The command line input for the generation of the die is given as

```
BLCK5,0.0,1.102,0.0,2.5,0.3,10,2,11
```

The mesh for the mold is achieved in three stages: (i) the mesh next to the die-attach, (ii) the mesh next to the die, and (iii) the mesh above everything. The commands for these steps are the same as those used before, so no details are given here. Figures 2.50-2.61 give the dialog boxes with appropriate parameters and the resulting mesh configurations. The command line input for the generation of the mold mesh next to the die-attach is given as

```
BLCK1,2.75,1.032,0.25,0.5,2,7,0.146,0.235,0.07,3,2,1,1,8,12,0
BLCK4,3.5,1.032,0.0,0.5,0.07,4,45,2,1,14,12
```

The command line input for the generation of the mold mesh next to the die is given as

```
BLCK1,2.75,1.102,0.25,0.5,2,7,0.146,0.235,0.3,3,2,1,2,8,12,0
BLCK4,3.5,1.102,0.0,0.5,0.3,4,45,2,2,14,12
```

The command line input for the generation of the mold mesh covering the top surface of the package is given as

```
BLCK1,2.75,1.402,0.25,0.5,2,7,0.146,0.235,0.36,3,2,1,2,8,12,0
BLCK4,3.5,1.402,0.0,0.5,0.36,4,45,2,2,14,12
BLCK5,0.0,1.402,0.0,2.5,0.36,10,2,12
```

Figure 2.62 shows the final status of the mesh. Since the global model is for a nonlinear solution, the user must change the element type for the solder from SOLID45 to VISCO107 using CHGVISC, as shown in Fig. 2.63. The command line input for this step is given as

```
CHGVISC,1
```

Once the material properties have been specified, the user is ready to proceed to the solution phase of the global model. The material property

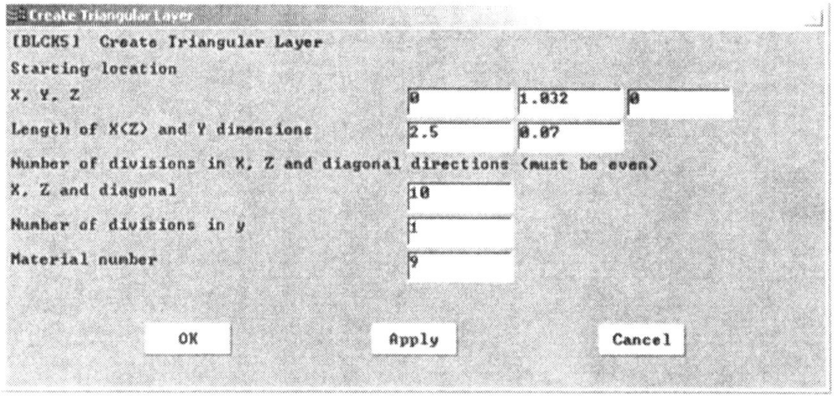

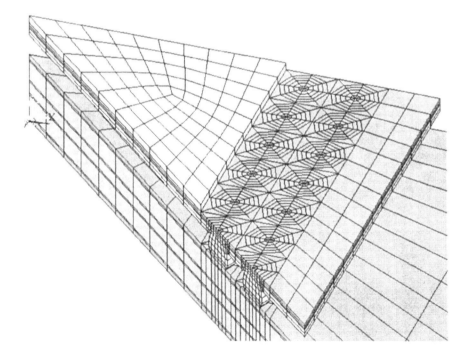

Create Triangular Layer

[BLCK5] Create Triangular Layer

Starting location

X, Y, Z |0 |1.032 |0

Length of X(Z) and Y dimensions |2.5 |0.07

Number of divisions in X, Z and diagonal directions (must be even)

X, Z and diagonal |10

Number of divisions in y |1

Material number |9

| OK | Apply | Cancel |

Fig. 2.46   Data used to create the die-attach mesh inside the solder joint area
(BLCK5 command).

Fig. 2.47   Oblique view of the mesh after the mesh for the die-attach is completed.

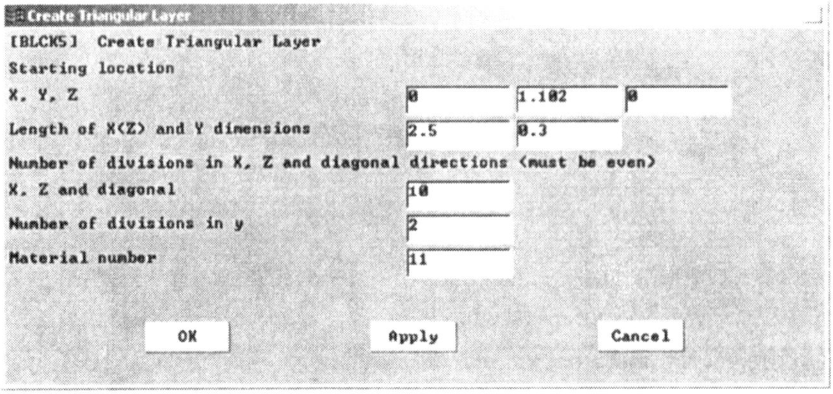

Create Triangular Layer

[BLCK5] Create Triangular Layer

Starting location

| X, Y, Z | 0 | 1.102 | 0 |

Length of X(Z) and Y dimensions   2.5   0.3

Number of divisions in X, Z and diagonal directions (must be even)

X, Z and diagonal   10

Number of divisions in y   2

Material number   11

| OK | Apply | Cancel |

Fig. 2.48   Data used to create the die mesh inside the solder joint area
(BLCK5 command).

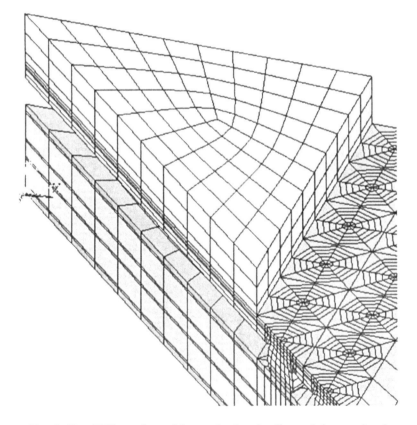

Fig. 2.49   Oblique view of the mesh after the die mesh is completed.

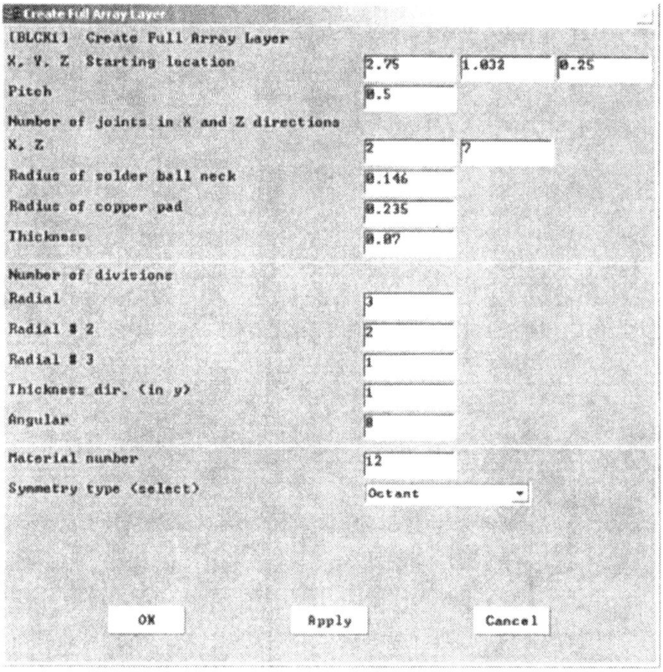

Create Full Array Layer

[BLCK1] Create Full Array Layer

X, Y, Z Starting location	2.75	1.032	0.25
Pitch	0.5		
Number of joints in X and Z directions			
X, Z	2	7	
Radius of solder ball neck	0.146		
Radius of copper pad	0.235		
Thickness	0.07		
Number of divisions			
Radial	3		
Radial # 2	2		
Radial # 3	1		
Thickness dir. (in y)	1		
Angular	8		
Material number	12		
Symmetry type (select)	Octant		

OK   Apply   Cancel

Fig. 2.50   Data used to create the mold mesh above the solder joint area (BLCK1 command).

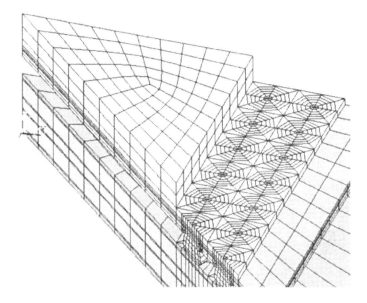

Fig. 2.51   Oblique view of the mesh after the BLCK1 command is executed.

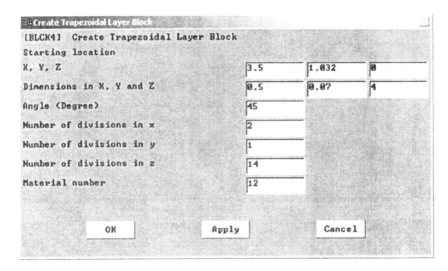

Fig. 2.52   Data used to create the mold mesh outside the solder joint area
(BLCK4 command).

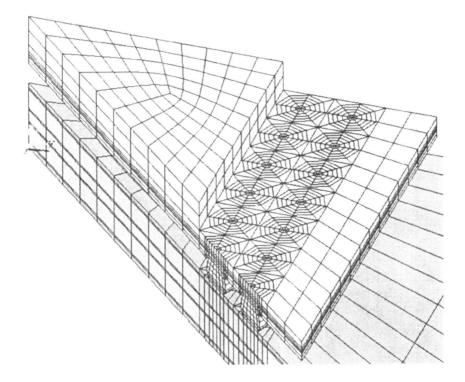

Fig. 2.53   Oblique view of the mesh after the first layer of the model is
completed.

```
Create Full Array Layer
[BLCK1]  Create Full Array Layer
X, Y, Z  Starting location           2.75        1.102       0.25
Pitch                                0.5
Number of joints in X and Z directions
X, Z                                 2           7
Radius of solder ball neck           0.146
Radius of copper pad                 0.235
Thickness                            0.3

Number of divisions
Radial                               3
Radial # 2                           2
Radial # 3                           1
Thickness dir. (in y)                2
Angular                              8

Material number                      12
Symmetry type (select)               Octant         ▾


            OK              Apply              Cancel
```

Fig. 2.54 Data used to create the mold mesh above the solder joint
area (BLCK1 command).

```
Create Trapezoidal Layer Block
[BLCK4]  Create Trapezoidal Layer Block
Starting location
X, Y, Z                              3.5         1.102       0
Dimensions in X, Y and Z            0.5          0.3         4
Angle (Degree)                       45
Number of divisions in x             2
Number of divisions in y             2
Number of divisions in z             14
Material number                      12


            OK              Apply              Cancel
```

Fig. 2.55 Data used to create the mold mesh outside the solder joint area
(BLCK4 command).

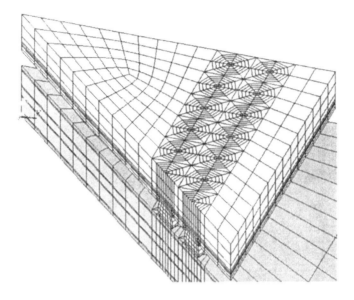

Fig. 2.56   Oblique view of the mesh after the second layer of the mold
mesh is completed.

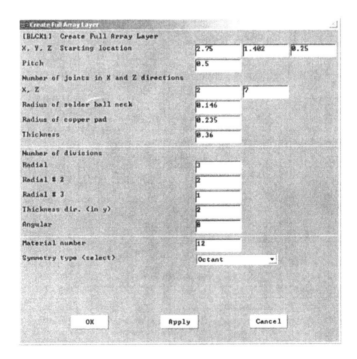

Fig. 2.57   Data used to create the mold mesh above the solder joint area
(BLCK1 command).

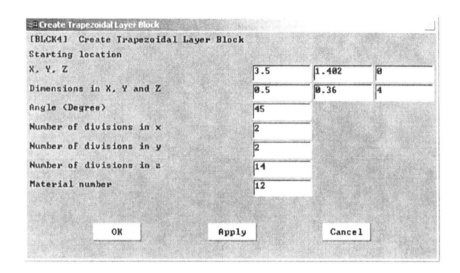

Fig. 2.58 Data used to create the mold mesh outside the solder joint area (BLCK4 command).

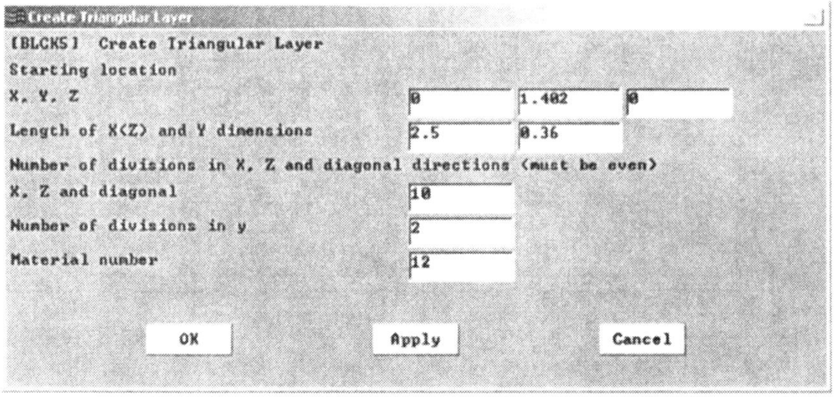

Fig. 2.59 Data used to create the mold mesh inside the solder joint area (BLCK5 command).

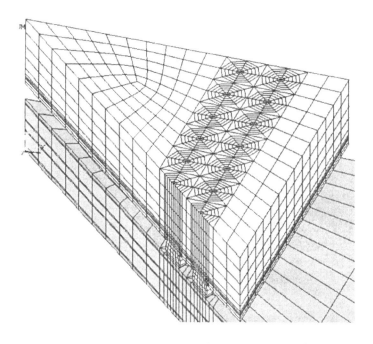

Fig. 2.60   Oblique view of the mesh after the mold mesh is completed.

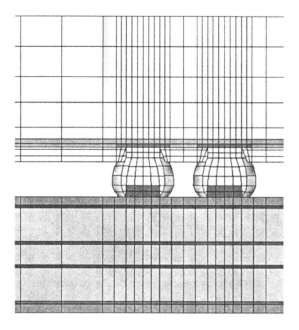

Fig. 2.61   Front view of the mesh after the mold mesh is completed.

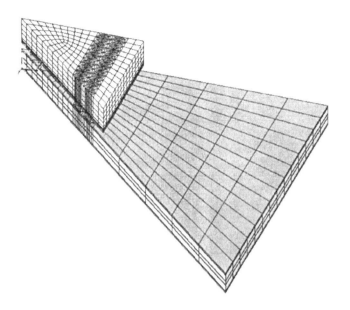

Fig. 2.62   Oblique view of the final mesh of the global model.

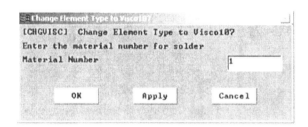

Fig. 2.63   Dialog box for the CHGVISC command.

correspondence is given in Table 2.1. Figure 2.64 shows the EPMAT dialog box for material #1 (in ANSYS®), with nonlinear solder material properties selected. The remaining material properties are selected similarly. The command line input for this step is given as

```
EPMAT,1,26
EPMAT,11,15
EPMAT,12,27
EPMAT,9,28
EPMAT,8,29
EPMAT,7,30
EPMAT,6,31
EPMAT,2,14
EPMAT,3,14
EPMAT,4,11
EPMAT,5,30
```

Table 2.1   Material property correspondence in
the thermomechanical case study.

Material	Material # in	
	ANSYS®	EPMAT Command
Solder	1	26
Copper	2	14
Copper	3	14
FR4	4	11
Polyimide	5	30
Adhesive	6	31
Substrate	7	30
Solder Mask	8	29
Die-attach	9	28
Die	11	15
Mold	12	27

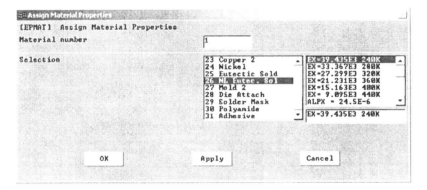

Fig. 2.64   Dialog box for the EPMAT command.

2.3.3 Global Model Solution and Preparation for Submodeling

In this section, the symmetry boundary conditions and displacement con-
straint on one node are specified first. Then, the thermal cycle simulation
load step files are created. Finally, the solution procedure is initiated. The
GL_1 command is used to specify symmetry boundary conditions, as shown
in Fig. 2.65. Rigid-body translation is prevented through use of the GL_2
command (Fig. 2.66). The command line input for these steps is given as

```
GL_1,0
GL_2,0,0,0
```

Finally, the load step files for the simulation of the thermal cycle are created
by using SUB5, as shown in Fig. 2.67. The command line input for this step
is given as

```
SUB5,4,373.15,233.15,298.15,0.0,1560,1440,240,360,6,6,1
```

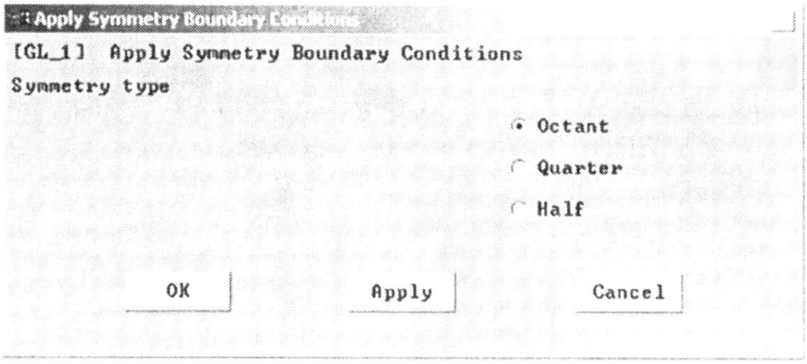

Fig. 2.65   Dialog box for the GL_1 command.

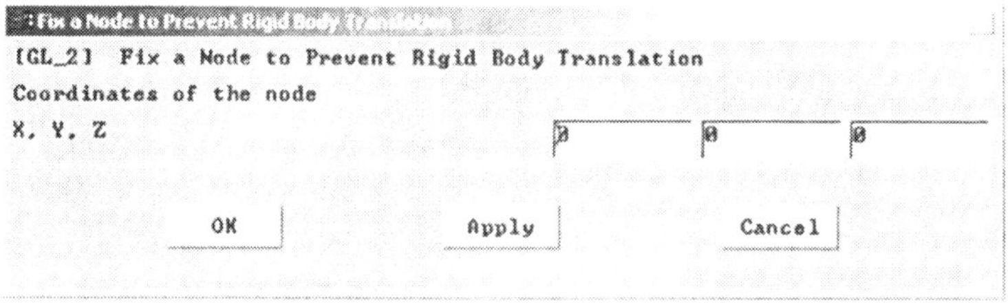

Fig. 2.66   Dialog box for the GL_2 command.

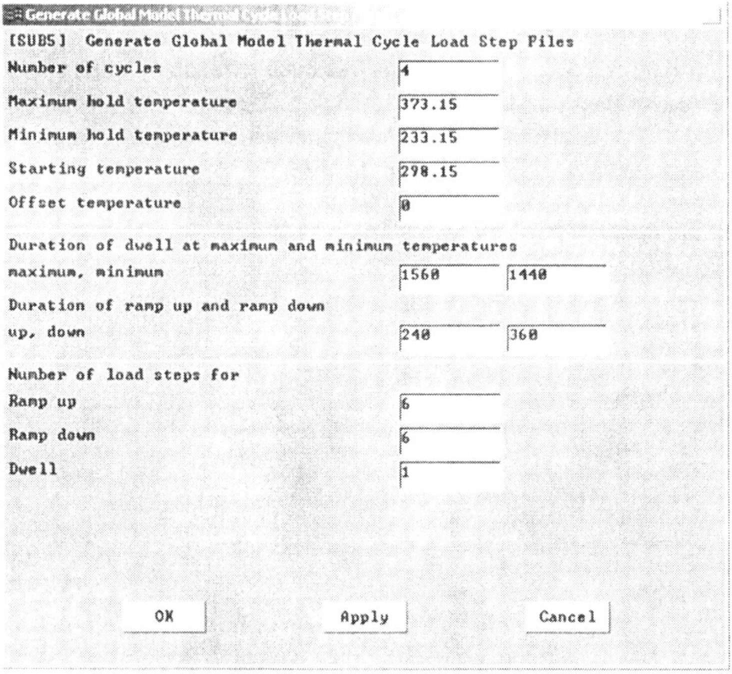

Fig. 2.67   Dialog box for the SUB5 command.

At this point, it is strongly recommended that the user select all the entities (select everything command in ANSYS®), **save** the work, and start the solution by using LSSOLVE, as shown in Fig. 2.68. The user **must** also save the work after the solution is done. The command line input for these steps is given as

```
allsel,all
SAVE
LSSOLVE,1,60
SAVE
```

Once the solution for the global model is complete, the user should evaluate the most critical solder joint. This step of the analysis requires the selection of the most critical solder joint, and several criteria for doing so have been presented in the literature. In this particular method (nonlinear global model with nonlinear submodel), the maximum value of the plastic work at the end of the last thermal cycle is used as the criterion. The contour plot of plastic work at the end of the fourth thermal cycle is shown in Fig. 2.69. The outer solder joint on the diagonal of the package is clearly the most critical solder joint. This step provides the user with information on the x- and z-coordinates of the center of the solder joint, which will be used in the generation of the submodel.

2.3.4 Submodel Generation

Once the most critical solder joint has been evaluated, as explained in the preceding section, the user should generate the mesh for the submodel. The submodel is a refined model of this joint and its vicinity. Figures 2.70 and 2.71 show the submodel created in this case.

Since the generation of the mesh involves similar, if not the same, steps, the details are not discussed here. However, the command line input is given

Fig. 2.68   Dialog box for the LSSOLVE command (global model).

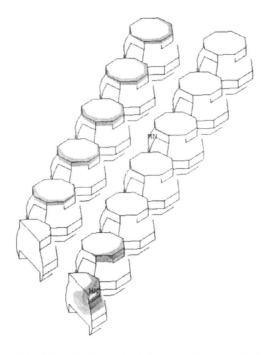

Fig. 2.69   Contour plot of the plastic work at the end of the fourth thermal cycle.

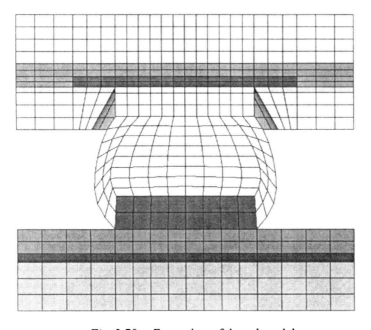

Fig. 2.70   Front view of the submodel.

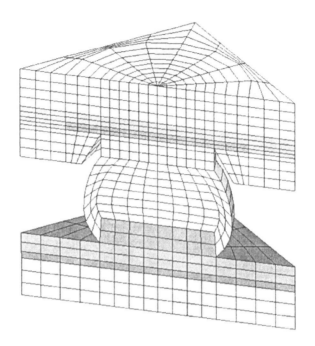

Fig. 2.71   Oblique view of the submodel.

below for the sake of completeness. Note that in SJFULL, the first and third arguments are the *x*- and *z*-coordinates of the center of the solder joint whose values were evaluated in the last step of the previous section.

```
!CREATE SOLDER BALL
SJGEO1,2,0.146,0.235,0.15,0.146,0.232,0.021,0.06,0.067,,1,0.01
    27,,
SJGEO2,7,0.174,0.029,0.188,0.058,0.192,0.087,0.19,0.116,0.185,
    0.145,0.177,0.174,0.165,0.203
SJGEO3,6,3,,2,2,2,,8,,16,,,,
SJFULL,3.25,0.692,3.25,0.5,1,1,1,2,0

!CREATE PCB
FR4GEO1,3,0.1,0.018,0.05
FR4GEO2,3,3,1,2
FR4GEO3,3,4,3,5
BLCK7,3.25,0.524,3.25,0.5,1,1,0.15,,4,,4,16,0

!CREATE ADHESIVE
BLCK2,3.25,0.972,3.25,0.5,1,1,0.146,0.235,0.012,60,3,3,1,16,
    6,0

!CREATE SUBSTRATE
PARAM1=0.146+0.012/TAN(ATAN(1)*4/3)
BLCK2,3.25,0.897,3.25,0.5,1,1,PARAM1,0.235,0.075,60,3,3,3,16,
    7,0
```

```
!CREATE SOLDER MASK
BLCK2,3.25,0.984,3.25,0.5,1,1,0.235,,0.021,90,6,3,2,16,8,0
BLCK1,3.25,1.005,3.25,0.5,1,1,0.146,0.235,0.027,6,3,3,2,16,8,0

!CREATE MOLD
BLCK1,3.25,1.032,3.25,0.5,1,1,0.146,0.235,0.1,6,3,3,4,16,12,0

!CHANGE SOLDER BALL ELEMENT TYPE TO VISCO107
CHGVISC,1

!SET MATERIAL PROPERTIES

!solder balls
EPMAT,1,26
!mold
EPMAT,12,27
!solder mask
EPMAT,8,29
!substrate
EPMAT,7,30
!adhesive
EPMAT,6,31
!copper pads
EPMAT,2,14
!motherboard
EPMAT,3,14
EPMAT,4,11
EPMAT,5,30
```

Once the submodel has been created, the user is ready to proceed to the solution phase. First, the interface nodes between the global model and the submodel must be identified and recorded in a file named "cutdof.node." This is done using CUTDOF, as shown in Fig. 2.72. The command line input for this step is given as

```
CUTDOF,0,0.524,1.132,3.0,3.5,3.0,3.25,0
```

At this point, the user **must** select everything and save the submodel. This is done using the command line input

```
allsel,all
SAVE
```

Next, the boundary conditions for each load step should be generated from the global model results using CBCYCL, as shown in Fig. 2.73. The command line input for this step is given as

```
CBCYCL,60,'DEMO','DEMOSUB'
```

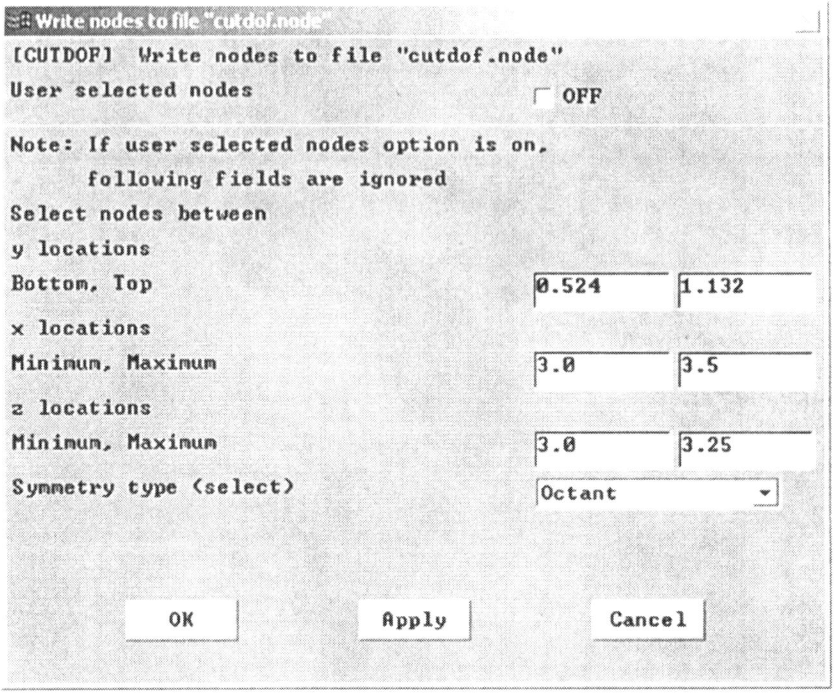

Fig. 2.72   Dialog box for the CUTDOF command.

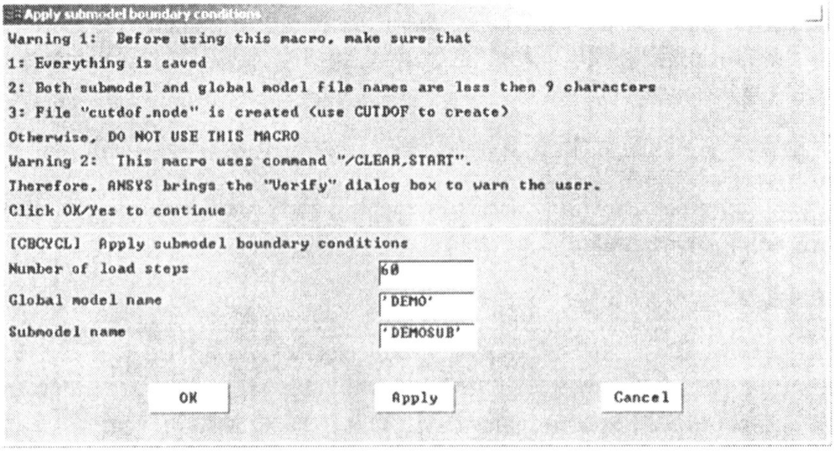

Fig. 2.73   Dialog box for the CBCYCL command.

As mentioned in Chap. 4, this command activates the /clear command twice, which clears the database, and ANSYS® requires confirmation each time. The user **must** hit OK to confirm.

After the boundary conditions for each load step are extracted from the global model solution, the load step files for the submodel must be created. This is done by use of SUB6. Figure 2.74 shows the parameters used for this command. The command line input for this step is given as

```
/SOLU
SUB6,4,373.15,233.15,298.15,0.0,1560,1440,240,360,6,6,1
```

Once the load step files have been created for the submodel, the user can start the solution procedure using LSSOLVE, as shown in Fig. 2.75. It is strongly recommended that the user save the work before starting the solution procedure. The user should save the work after the solution, as well. The command line input for these steps is given as

```
allsel,all
SAVE
LSSOLVE,1,60
SAVE
FINISH
```

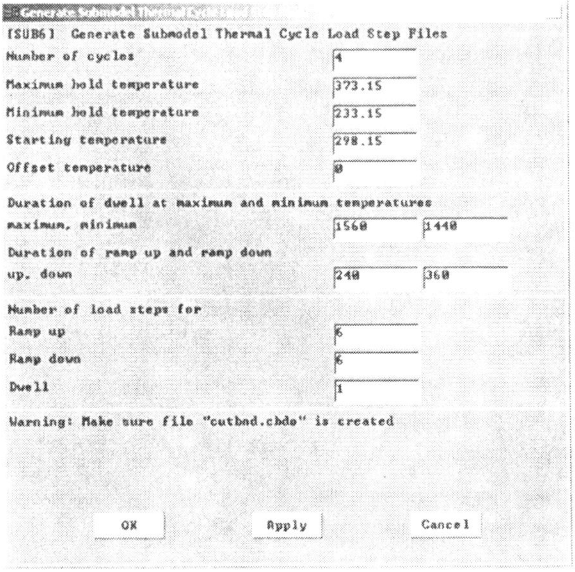

Fig. 2.74   Dialog box for the SUB6 command.

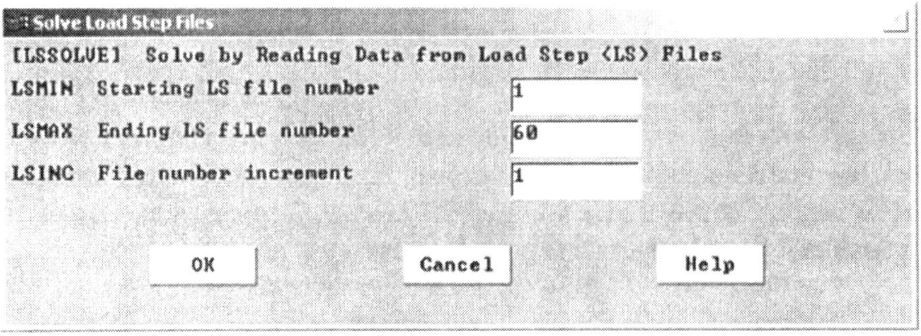

Solve Load Step Files

[LSSOLVE] Solve by Reading Data from Load Step (LS) Files

LSMIN Starting LS file number   1

LSMAX Ending LS file number   60

LSINC File number increment   1

| OK | Cancel | Help |

Fig. 2.75   Dialog box for the LSSOLVE command (submodel).

2.3.5 Submodel Solution and Life Calculation

Once the solution for the submodel has been obtained, a contour plot of the plastic work at the end of the fourth cycle is reviewed, as shown in Fig. 2.76. It is observed that the package side is the critical side and therefore the life prediction will be performed for this side.

The predicted life is calculated by using LIFE1 under General Post processing > Packaging. The dialog box for this is given in Fig. 2.77. This operation will result in the calculation of the predicted life and the volume-weighted-average plastic work for the selected elements during the fourth cycle, as shown in Fig. 2.78.

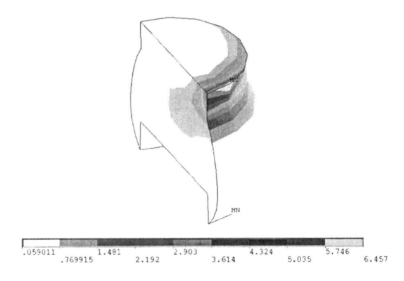

| .059011 | 1.481 | 2.903 | 4.324 | 5.746 |
| .769915 | 2.192 | 3.614 | 5.035 | 6.457 |

Fig. 2.76   Contour plot of the plastic work in the submodel.

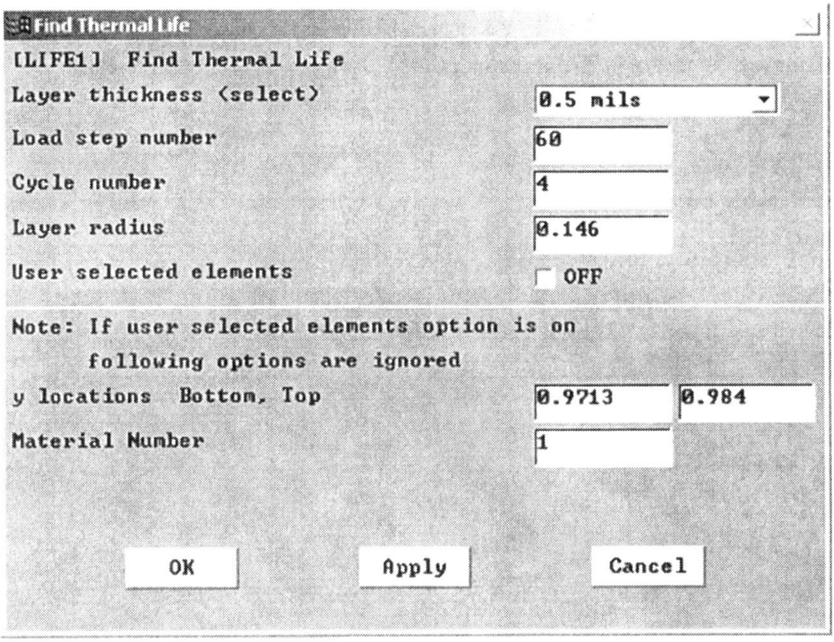

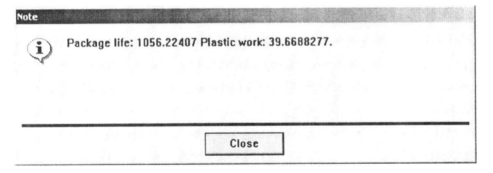

Fig. 2.77 Dialog box for the LIFE1 command.

Fig. 2.78 Pop-up window in ANSYS® reporting the values of the predicted package life and volume-weighted-average plastic work.

2.4 References

Darveaux, R., 1997, "Solder Joint Fatigue Life Model," *Design and Reliability of Solder and Solder Interconnections, Proceedings of the TMS*, The Minerals, Metals, and Materials Society, Warrendale, pp. 213-218.

Gustafsson, G., Guven, I., Kradinov, V., and Madenci, E., 2000, "Finite Element Modeling of BGA Packages for Life Prediction," *Proceedings, 50th Electronic Components and Technology Conference,* IEEE, New York, pp. 1059-1063.

Riebling, J., 1996, "Finite Element Modeling of Ball Grid Array Components," Master's thesis, State University of New York at Binghamton, Binghamton, NY.

Chapter 3

MECHANICAL BENDING FATIGUE LIFE
PREDICTION ANALYSIS

The fatigue life prediction of solder joints under cyclic bending requires a three-stage analysis: (1) a three-dimensional finite element analysis with linear material properties for computing the assembly stiffness and imposed strain values, (2) a one-dimensional combined creep and time-independent plastic deformation analysis under cyclic loading for computing the inelastic strain energy, and (3) a life prediction analysis.

The response of a solder joint under cyclic bending reaches the steady-state response after a large number of cycles. Therefore, the three-dimensional nonlinear finite element analysis simulation of a solder joint until achieving a stable hysteresis loop (steady-state response) is computationally impractical. In order to render the analysis tractable, the mechanical bending fatigue analysis is based on the concept of the assembly stiffness of the most critical solder joint introduced by Darveaux and Banerji (1991).

3.1 Approach

This approach assumes that the most critical solder joint experiences a reduction in stiffness due to the presence of a damaged zone near either the board side or the component side, as shown in Fig. 3.1. The effective stiffness of the damaged zone is represented by S_e^ℓ, and the assembly stiffness for the entire solder, S, is obtained by a series combination of the stiffnesses of the damaged and the undamaged portions of the solder joint as

$$\frac{1}{S} = \frac{1}{S_e^\ell} + \frac{1}{E} \tag{3.1}$$

where E represents the Young's modulus of the solder alloy.

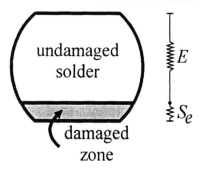

Fig. 3.1   Series combination of the stiffnesses representing the
damaged and undamaged zones of the solder ball.

The effective stiffness calculation of the damaged zone of the most critical
solder joint requires a three-dimensional linear finite element analysis of the
electronic package, first with *undamaged* solder joints and then with *damaged* solder joints, both under the specified maximum load arising from
bending. The Young's modulus of the damaged solder joint is estimated by
reducing the modulus of the undamaged joint by a factor of 100, as suggested by Darveaux (2000). The most critical solder joint is established
based on the equivalent stress values within the solder joints, as suggested
by Darveaux and Syed (2000). The effective stiffness and imposed strain
values are computed by volumetrically averaging equivalent stress and
strain values for the finite elements near the top and bottom of the most
critical solder joint. For the solution with the undamaged solder modulus,
the volume-averaged equivalent (von Mises) stress and strain values are
represented by σ_1^ℓ and ε_1^ℓ, respectively. The superscript ℓ designates the
location of the damage zone, i.e., component or board side of the solder
joint. Similarly, $\sigma_{0.01}^\ell$ and $\varepsilon_{0.01}^\ell$ represent the volume-averaged equivalent
stress and strain values, respectively, for the solution with the modulus of
the damaged solder.

Corresponding to the damaged and undamaged solder joints, these volume-averaged equivalent (von Mises) stress and strain values permit the determination of the effective stiffness, S_e^ℓ, and imposed strain, ε_0^ℓ, of the most
critical solder joint. As shown in Fig. 3.2, they are computed as

$$S_e^\ell = -(\sigma_1^\ell - \sigma_{0.01}^\ell)/(\varepsilon_1^\ell - \varepsilon_{0.01}^\ell), \quad \varepsilon_0^\ell = \varepsilon_{0.01}^\ell + \frac{\sigma_{0.01}^\ell}{S_e^\ell} \qquad (3.2)$$

with the superscript ℓ designating the damage site (component or board
side).

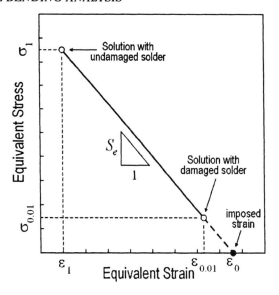

Fig. 3.2   Effective stiffness and imposed strain calculation.

Note that the empirical constants of the life predition model are dependent on the element thickness. Therefore, the finite element discretization of the solder ball should be consistent with that of suggested by Darveaux and Syed (2000). If the joint interface is of the solder mask defined (SMD) type, two layers of elements, each 0.0005-inch thick, are suggested in the volumetric averaging. If the joint type is non-SMD (NSMD), as shown in Fig. 3.3, the fillet layer around the pad and two 0.0005-inch-thick layers of elements are suggested.

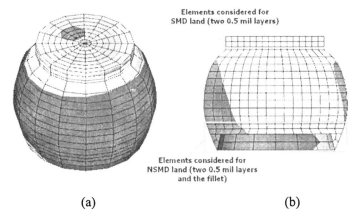

(a)   (b)

Fig.3.3   Equivalent stress contour plot of the critical joint (a) and the elements considered in volume-weighted averaging for SMD and NSMD types of lands (b).

Characterization of the deformation behavior of the most critical solder joint based on the assembly stiffness, S, permits a one-dimensional nonlinear analysis under cyclic loading until a stable (steady-state) response is achieved. In the one-dimensional analysis, the applied strain, ε_t, corresponding to the mechanical cyclic loading is periodic in the range of $[0, \varepsilon_0]$, and it is represented in the form of a function as

$$\varepsilon_t = \varepsilon_0 \sin^2 \frac{\pi t}{P} \qquad (3.3)$$

in which P is the period of loading and t denotes time. As an alternative to this simple periodic function, the following form

$$\varepsilon_t = \varepsilon_0 - \varepsilon_0 \left(\frac{2t - P}{P}\right)^6 H(1-t) + \varepsilon_0 \left(2\frac{t-P}{P}\right)^6 H(t-1) \qquad (3.4)$$

may also be considered, with $H(t)$ representing the Heaviside step function. Although these two function forms are drastically different in behavior within each period, the difference in the strain energy increase per cycle is negligible. Therefore, a simple sinusoidal function is used to represent the applied loading.

3.1.1 Combined Creep and Time-Independent Plastic Deformation

For a specified assembly stiffness, imposed strain, and period of cyclic loading, a one-dimensional nonlinear deformation of the solder can be obtained by solving the coupled nonlinear ordinary differential equations given in Eqs. (1.10), (1.11), and (1.13)-(1.15) through a numerical algorithm utilizing Muller's method. These equations are in the form

$$\frac{\sigma}{S} + C_{6t} \left(\frac{\sigma}{G}\right)^m + \varepsilon_c - \varepsilon_t = 0 \qquad \text{loading} \quad (3.5)$$

$$\frac{\sigma}{S} + \tilde{\varepsilon}_p - \frac{C_{6t}}{G^m} 2^{1-m} (\tilde{\sigma} - \sigma)^m + \varepsilon_c - \varepsilon_t = 0 \qquad \text{unloading} \quad (3.6)$$

$$\frac{\sigma}{S} + \bar{\varepsilon}_p + \frac{C_{6t}}{G^m} 2^{1-m} (\sigma - \bar{\sigma})^m + \varepsilon_c - \varepsilon_t = 0 \qquad \text{reloading} \quad (3.7)$$

in which the creep strain, ε_c, is obtained by integration of

$$\dot{\varepsilon}_c = \dot{\varepsilon}_{ss} \left(1 + \varepsilon_T B_T e^{-B_T \dot{\varepsilon}_{ss} t}\right) \quad \text{with} \quad \dot{\varepsilon}_{ss} = C_{5t} \left[\sinh(\alpha_{1t}\sigma)\right]^n e^{-Q_a/kT}$$

As a result of numerical integration, the creep strain, plastic strain, effective stress, and inelastic and total strain energies are calculated at every time step for a specified number of cycles. In order to achieve the steady-state response, a sufficient number of cycles must be employed in the simulation. Associated with the stable hysteresis loop (the steady-state response), the computed inelastic and total strain energies can be used for life prediction of the package utilizing the criterion suggested by Darveaux (2000), provided that the empirical constants are based on a combined creep and time-independent plasticity analysis.

3.1.2 Life Prediction

Life prediction of a package can be based on the relationship given by Eq. (1.18), with the empirical constants provided by Darveaux and Syed (2000). Their values are presented in Table 1.8.

In computing the strain energy, the total strain (sinusoidal function of time) is applied in small increments, and the corresponding stress, σ_n, is computed at the end of each time step, leading to the calculation of the strain energy by trapezoidal integration. The strain energy increase in a cycle is calculated as the difference between strain energy at the beginning and ending of that load cycle.

3.2 Analysis Steps

This section describes the analysis steps for creating a *linear global model with one-dimensional combined creep and time-independent plasticity*.

As in the case of the thermomechanical analysis, the user must have detailed geometric information about the actual package for an accurate description of the local geometry of the solder joints. The necessary geometrical information includes the pitch, copper pad radii and thicknesses, standoff height, and all dimensions (length, width, and thickness) of every component in the package. In the absence of symmetry in the loading, the user must consider the entire package in the model generation and in the solution stages of the analysis.

The analysis starts with the model generation, followed by the simulation of the mechanical loading representing bending, determination of the assembly stiffness, and a one-dimensional analysis for fatigue life calculation. Although the problem has time dependence, only a static solution corresponding to the maximum load is obtained in the three-dimensional finite element method analysis stage. The basic steps of the analysis are given below.

- Model generation.
- Use linear-elastic material properties for the solder.
- Apply symmetry conditions if applicable.
- Apply displacement constraints on nodes that correspond to the support locations.
- Apply appropriate displacement constraints to prevent rigid-body displacements.
- Apply force boundary conditions (maximum load) at nodes that correspond to the location of the applied load.
- Solve.
- Identify the most critical solder joint.
- Select elements at a possible crack initiation site.
- Obtain and store volume-weighted-average von Mises stress and strain (σ_1^ℓ and ε_1^ℓ) for the selected elements.
- Reduce Young's modulus for the solder by a factor of 100.
- Solve under same boundary conditions.
- Select elements at a possible crack initiation site (same elements as before).
- Obtain and store volume-weighted-average von Mises stress and strain ($\sigma_{0.01}^\ell$ and $\varepsilon_{0.01}^\ell$) for the selected elements at the same solder joint.
- Find the effective stiffness, imposed strain and, subsequently, assembly stiffness.
- Perform one-dimensional combined creep and time-independent plasticity simulation.
- Calculate predicted life using Eq. (1.18).

ReliANS, a group of ANSYS® macros, was designed/developed to carry out these analysis steps in a seamless fashion within the ANSYS®-GUI environment. These macros, not available in the standard ANSYS® program, must be installed by following the instructions provided in Appendix A.

3.3 Case Study: BGA-Type Package

3.3.1 Data Needed Before Beginning ANSYS® Session

A demonstration of the use of the mechanical bending fatigue life prediction analysis using ReliANS commands is given by considering a BGA-type package with 132 solder balls arranged in a three-row peripheral pattern. As shown in Fig. 3.4, there are 14 solder balls in each direction (*x* and *z*) along the outermost ring. All the dimensions are given in millimeters unless otherwise specified. The pitch is 0.8 mm. Die and substrate sizes are given as 8.0 × 8.0 × 0.3 mm^3 and 12.0 × 12.0 × 0.075 mm^3, respectively. A cross-sectional view is illustrated in Fig. 3.5.

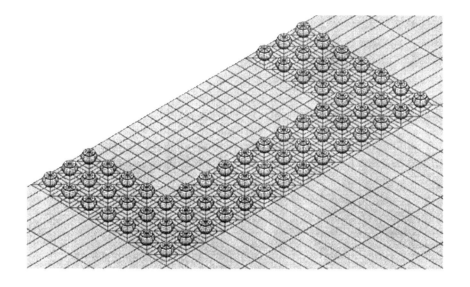

Fig. 3.4   Solder joint pattern.

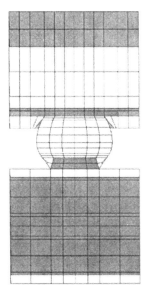

Fig. 3.5   Cross-section of the package.

The package is subjected to a cyclic loading at its center. The loading and unloading range varies between 0 and 30 N over a period of 1 sec. The three support points, located on a circle of 12-mm radius, form an equidistant triangle, and the origin of the circle coincides with the center of the package. The dimensions of the PCB are taken as $32.5 \times 32.5 \text{ mm}^2$, with a thickness of 0.692 mm. Having all the geometrical and material information available, one can start the modeling phase with the global model.

3.3.2 Global Model Generation

Due to the presence of symmetry in loading and support conditions, it suffices to consider only one half of the package in the analysis, as shown in Fig. 3.6.

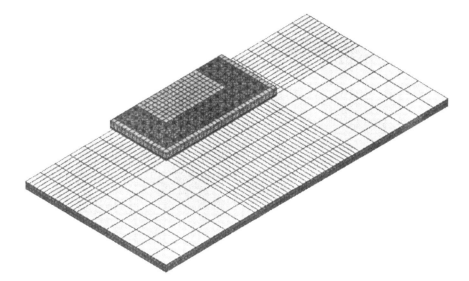

Fig. 3.6   Package with half symmetry.

The maximum load is applied by uniformly distributing it over a 1-mm-radius area. Because this analysis step is identical to the one solved for the thermomechanical life prediction analysis, the intermediate steps of global model generation are not discussed in this context. However, an input file that contains the commands to create the global model for this package is included in Appendix B.

The displacement constraints corresponding to the supports during the cyclic bending load are specified using GLMEC1, for which the dialog box is shown in Fig. 3.7. Similarly, the maximum load of 30 N is applied using SUB20, as shown in Fig. 3.8.

3.3.3 Global Model Solution and Preparation for Creep/Plasticity Simulation

The most critical solder joint is established based on the equivalent stress values among all the solder joints, as suggested by Darveaux and Syed (2000). Examination of the von Mises stresses in the solder balls reveals that the most critical solder ball is located at the bottom corner. The effective stiffness, imposed strain, and the assembly stiffness values are then calculated by volumetrically averaging equivalent stress and strain values for the elements near the top and bottom of the most critical solder joint.

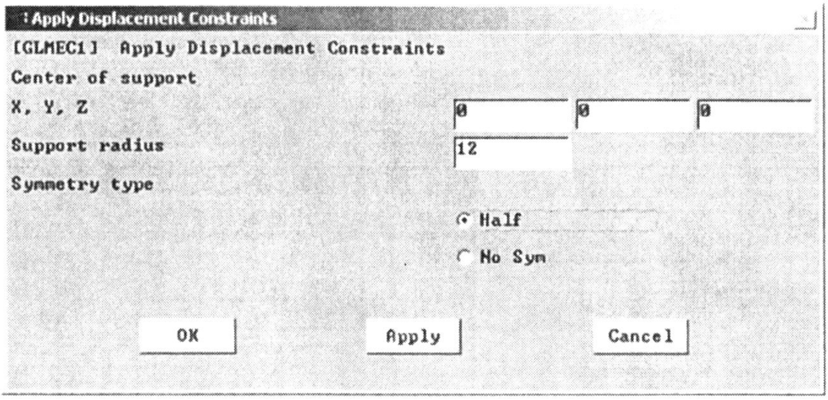

Fig. 3.7   Command used for the application of the support condition.

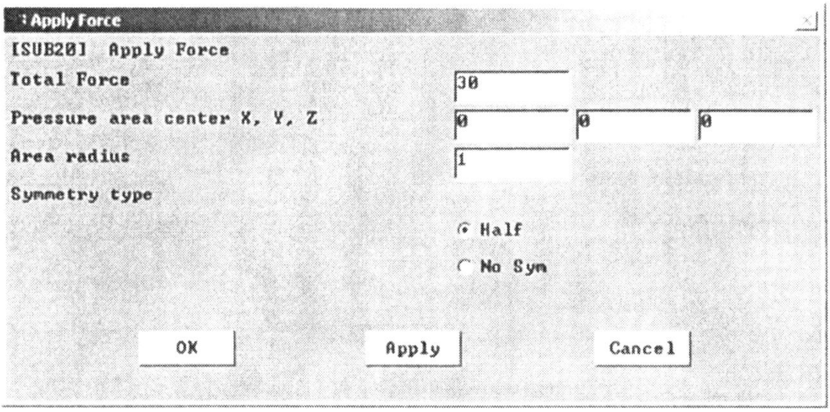

Fig. 3.8   Command used for the application of the force exerted by the load cell.

The volume-weighted-average von Mises stress and strains are calculated using F_VONM (as shown in Fig. 3.9).

For the component side (top):

$$\sigma_1^\ell = 69.4872\,\text{MPa} \qquad \sigma_{0.01}^\ell = 11.6367\,\text{MPa}$$
$$\varepsilon_1^\ell = 0.0031 \qquad \varepsilon_{0.01}^\ell = 0.0513$$

For the board side (bottom):

$$\sigma_1^\ell = 65.2899\,\text{MPa} \qquad \sigma_{0.01}^\ell = 11.2215\,\text{MPa}$$
$$\varepsilon_1^\ell = 0.0029 \qquad \varepsilon_{0.01}^\ell = 0.0495$$

The corresponding values for the effective stiffness, imposed strain, and assembly stiffness are calculated using the F_E_EFF command (as shown in Figs. 3.10 and 3.11).

For the component side (top):

$$S_e^\ell = 173891.1362\,\text{MPa}, \quad \varepsilon_0 = 0.06102159, \quad S = 167337.6828\,\text{MPa}$$

For the board side (bottom):

$$S_e^\ell = 168262.2321\,\text{MPa}, \quad \varepsilon_0 = 0.05915735, \quad S = 162118.6912\,\text{MP}$$

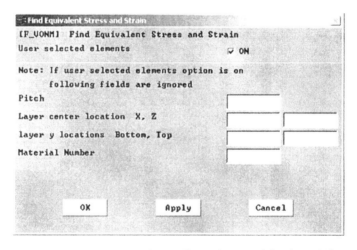

Fig. 3.9   Command used for finding volume-weighted von Mises stress and strain for the selected elements.

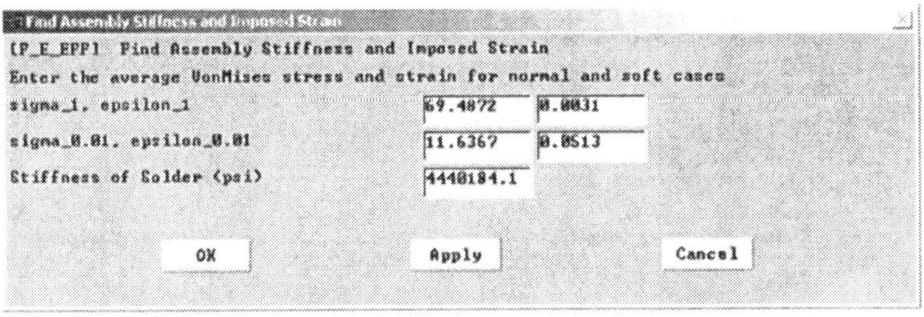

Fig. 3.10 Command for finding the assembly stiffness–component side.

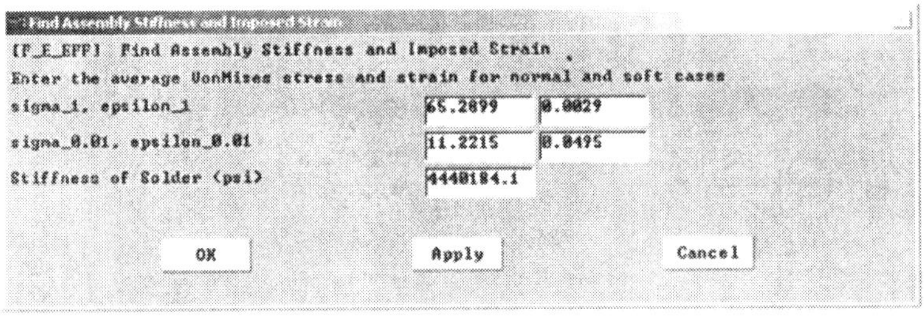

Fig. 3.11 Command for finding the assembly stiffness–board side.

These results are reported to the user by the information windows, as shown in Fig. 3.12 (for the component side) and Fig. 3.13 (for the board side).

3.3.4 Combined Creep and Plasticity Simulation

In order to employ the empirical constants provided by Darveaux and Syed (2000), the life prediction is based on an analysis of one-dimensional combined creep with time-independent plasticity. The total strain (sinusoidal function of time) is applied in small increments. In this case, each load cycle is divided into 100 time steps, i.e., each step is 0.01 sec. It is observed that the time-independent plasticity is much more dominant than the creep. Therefore, the strain energy increment per cycle does not exhibit a significant increase after the first cycle, and it stabilizes in a few cycles thereafter. In this case, 10 cycles are simulated and the resulting total strain energy increment per cycle is calculated as

For the component side (top):

$$\Delta W_{tot} = 33.4326 \text{ psi}$$

For the board side (bottom):

$$\Delta W_{tot} = 25.0143 \text{ psi}$$

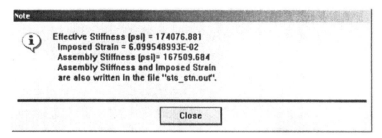

Fig. 3.12   ANSYS® window summarizing the assembly stiffness results–
component side.

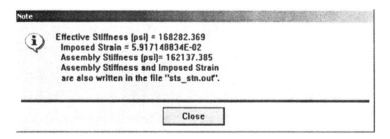

Fig. 3.13   ANSYS® window summarizing the assembly stiffness results–
board side.

These strain energy increments are obtained by using the F_BCLIFE command, as shown in Figs. 3.14 and 3.15 for the component and board sides, respectively. After the execution of this macro, an information window appears with information on ΔW_{tot} per cycle, number of cycles to crack initiation, and the characteristic life. For this case study, the information windows for the component and board sides are shown, respectively, in Figs. 3.16 and 3.17.

As apparent from the values of the strain energy increments, the component side is more critical than the board side. Using these values of the accumulated strain energy per cycle, the characteristic life for this package is calculated from Eq. (1.18).

For the component side (top):

$$N = 2392 \text{ cycles}$$

For the board side (bottom):

$$N = 3128 \text{ cycles}$$

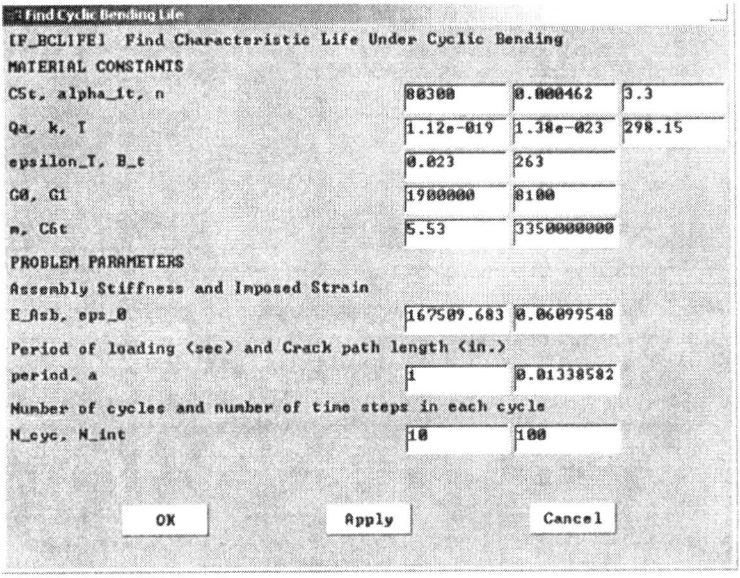

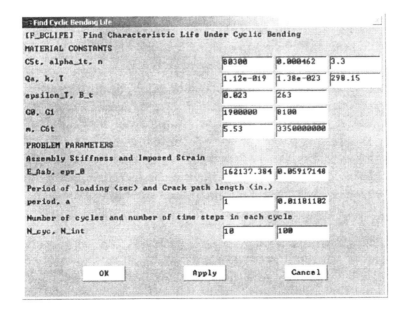

Fig. 3.14   Command for one-dimensional simulation and life prediction–
component side.

Fig. 3.15   Command for one-dimensional simulation and life prediction–
board side.

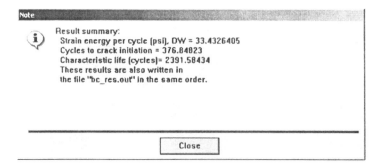

Fig. 3.16   ANSYS® window summarizing the life predition results–
component side.

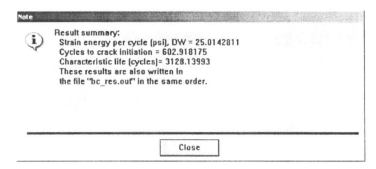

Fig. 3.17   ANSYS® window summarizing the life predition results–
board side.

The numerical values of the accumulated strain energy per cycle for the board and component sides are also written in ASCII format in the file "totwk.out" in the working directory. Also, the file "works.dat" stores the numerical values of ΔW_{tot} per cycle for determining if a stable solution is achieved. In addition, the file "result.dat" stores the time step number, time, stress, elastic strain, plastic strain, creep strain increment, creep strain, and strain energy at each time step. These files are overwritten each time the F_BCLIFE command is executed.

Using the information written in the "results.dat" file, the time-dependent behaviors of the stress, creep strain, plastic strain, and inelastic strain are examined in Figs. 3.18-3.21 for both the component and board sides. In all of the following figures, only the first four cycles are presented for the sake of clarity. Also, the variations of the stress as a function of the inelastic strain and total strain are shown in Figs. 3.22 and 3.23, respectively. These figures clearly indicate the stabilization of the hysteresis loop.

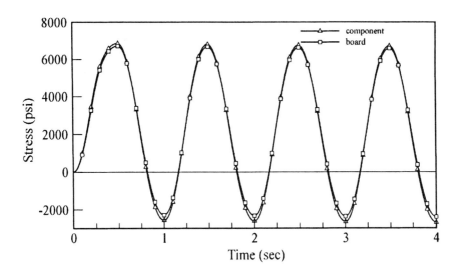

Fig. 3.18   Variation of stress with time over four cycles.

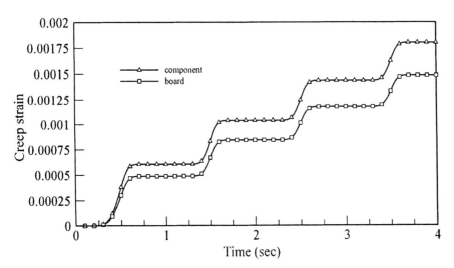

Fig. 3.19   Variation of creep strain with time over four cycles.

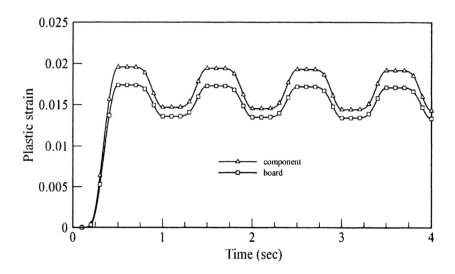

Fig. 3.20   Variation of plastic strain with time over four cycles.

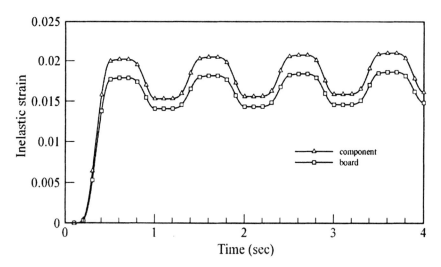

Fig. 3.21   Variation of inelastic strain with time over four cycles.

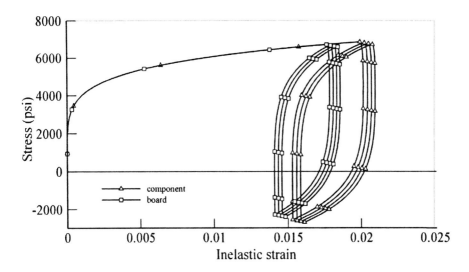

Fig. 3.22   Variation of stress with inelastic strain over four cycles.

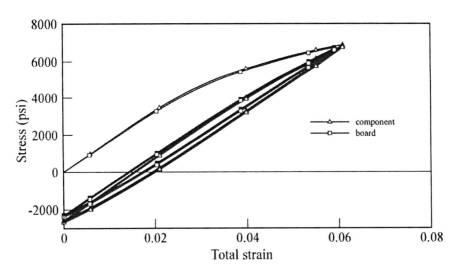

Fig. 3.23   Variation of stress with total strain over four cycles.

3.4 References

Darveaux, R., 2000, "Effect of Simulation Methodology on Solder Joint Crack Growth Correlation," *Proceedings, 50th Electronic Components and Technology Conference*, IEEE, New York, pp. 1048-1058.

Darveaux, R. and Banerji, K., 1991, "Fatigue Analysis of Flip Chip Assemblies Using Thermal Stress Simulations and a Coffin-Manson Relation," *Proceedings, 41st IEEE Electronic Components and Technology Conference*, IEEE, New York, pp. 797-805.

Darveaux, R. and Syed, A., 2000, "Reliability of Area Array Solder Joints in Bending," *SMTA International Proceedings 2000*, Surface Mount Technology Association, Edina, MN, pp. 313-324.

Chapter 4

MACRO REFERENCE LIBRARY

4.1 Overview

In this chapter, the usage of macros (user-defined commands) included in the add-on software is explained, along with descriptions of each individual parameter. A menu item named "*Packaging*" is added in the *preprocessor, solution,* and *postprocessor* menus within the ANSYS®-GUI environment, as shown in Fig. 4.1.

Under the *preprocessor* menu, the *packaging* menu is concerned with model generation and material properties. Under the *solution menu*, the *packaging* menu contains items related to the application of boundary conditions and the simulation of thermal and mechanical cycles. Finally, under the *post-processing* menu, the add-on menu item is used for the calculation of the expected life of the critical solder joint. The macro groupings are explained in the following subsections, starting with the *preprocessor* and followed by the *solution* and *postprocessing* menus.

4.2 Preprocessor

The *Packaging* menu contains six submenus:

Solder Joint	Geometrical parameters related to the solder joint are specified using this submenu. Also, a single or an array of solder joint meshes is generated under this submenu.
Motherboard	Geometrical parameters defining the layered structure of the printed circuit board (PCB) are specified under this submenu. Mesh generation macros (commands) include the PCB segment with an array of solder joint mesh patterns: triangular, rectangular, trapezoidal, or general block.

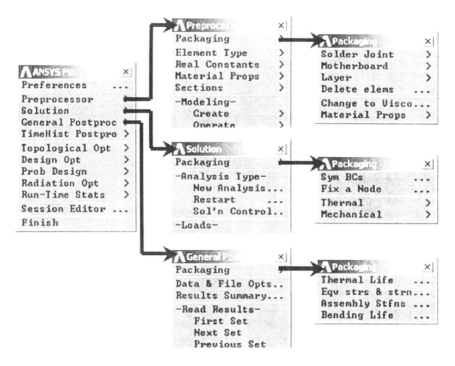

Fig. 4.1   Menu structure of the ReliANS add-on package.

Layer	This submenu is intended for the generation of meshes of the components of the package other than the solder joints and PCB, such as the die, die attach, solder mask, encapsulant, heat spreader, etc.
Delete Elems...	This menu item is used for the deletion of the last set of created elements. It is intended to serve as compensation for the lack of an *undo* function in ANSYS®.
Change to Visco...	When creating meshes for solder joints, ReliANS uses the SOLID45 element type, which is a brick element with linear material properties. However, before the solution is initiated, the element type for the solder elements should be changed to reflect the viscoplastic character of the material. This is achieved by executing this menu item.

Material Props   This submenu permits the user to pick and choose material properties from the database provided by ReliANS. These properties are a small selection of commonly used material properties for typical package components. They are given in two different measurement systems, one with megaPascals (MPa) and millimeters (mm) and the other with pounds per square inch (psi) and inches (in).

The structure of the *Packaging* menu is shown in Fig. 4.2, with submenus of each of the menu items given above. In Figs. 4.3-4.6, the *Solder Joint*, *Motherboard*, *Layer*, and *Material Props* submenus are shown, along with the equivalent commands of their menu items. Detailed explanations of each of the menu items (commands) are given below in the order given in Figs. 4.3-4.6.

IMPORTANT NOTE: Model generation with ReliANS utilizes ANSYS® solid modeling, thus creating keypoints, lines, areas, and volumes. The user must start with a *new session* when generating a new model using ReliANS. However, the user can resume with a "db" file previously generated by ReliANS.

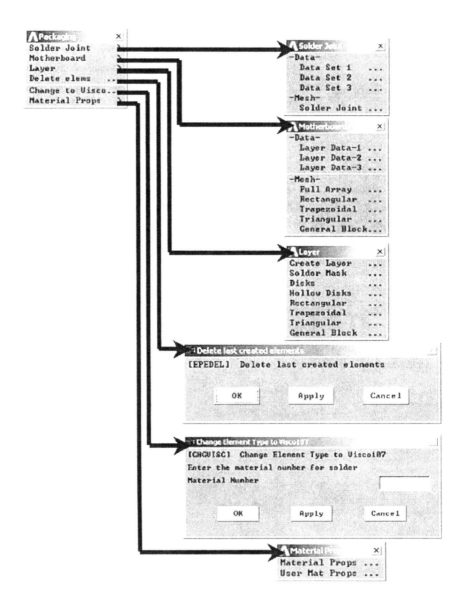

Fig. 4.2   Submenus of the *Packaging* menu under the *Preprocessor* menu.

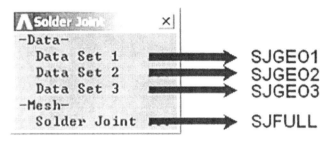

Fig. 4.3 Menu items and their equivalent commands for the *Solder Joint* submenu.

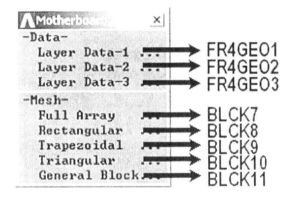

Fig. 4.4 Menu items and their equivalent commands for the *Motherboard* submenu.

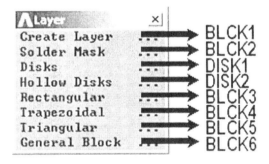

Fig. 4.5 Menu items and their equivalent commands for the *Layer* submenu.

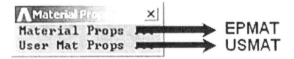

Fig. 4.6 Menu items and their equivalent commands for the *Material Props* submenu.

4.2.1 Solder Joint Submenu Commands

SJGEO1, *jtyp, tr, tr2, br, br2, h, tcu, tn, bcu, bn, nht, ht, nbt, bt*

This is the first of three data sets describing the geometry and meshing preferences of the solder joints. The definitions of the parameters are given
below, as well as in Figs. 4.7-4.11.[1] Figure 4.12 shows the dialog box for
this command.

jtyp	solder joint type (top/bottom)
	1: SMD/SMD
	2: SMD/NSMD
	3: NSMD/SMD
	4: NSMD/NSMD
	SMD: Solder Mask Defined
	NSMD: Non-Solder Mask Defined
tr	top joint radius
tr2	second top radius
br	bottom joint radius
br2	second bottom radius
h	height of the solder body; the user MUST refer to the figures for definitions of different joint types
tcu	top copper pad thickness
tn	top solder layer (neck) thickness (if SMD)
bcu	bottom copper pad thickness
bn	bottom solder layer (neck) thickness (if SMD)
nht	number of thin element layers on the top—see Notes
ht	thickness of each layer for the thin element layers on the top—see Notes
nbt	number of thin element layers on the top—see Notes
bt	thickness of each layer for the thin element layers on the top—see Notes

The menu path for this command is as follows:

```
Preprocessor > Packaging > Solder Joint > -Data- Data Set 1
```

Notes: The joint type parameter (*jtyp*) considers the top of the joint first
and the bottom second. Figures 4.8-4.11 demonstrate SMD/SMD, SMD/
NSMD, NSMD/SMD, and NSMD/NSMD, respectively. Since the plastic
work density values calculated at critical locations depend on the mesh
density around those regions, the prediction method that is currently used
(Darveaux 2000) provides different sets of empirical constants corre-

---

[1]In the rest of this chapter, the figures are given *after* the complete description of
each command (not necessarily on the page after they are mentioned).

sponding to different mesh densities. These constants are given for three different element thickness values (0.5 mil, 1.0 mil, and 1.5 mils). For this reason, the capability of having one of these predefined element thicknesses in the model is included in ReliANS. The user can place as many layers of elements (***nht*** and/or ***nbt***) as desired with any prescribed thickness (***ht*** and/or ***bt***) at the top or bottom of the solder joint. Refer to Figs. 4.8-4.11 for clarification.

The joint types and corresponding geometrical parameters given in Figs. 4.8-4.11 are for the most-general case. If, for example, the actual geometry of the top of the joint is similar to SMD but does not have a neck or a copper pad, the desired mesh can be obtained by simply entering zero values (or blank) for ***tr2***, ***tcu***, and ***tn***. However, this flexibility does not apply to NSMD types.

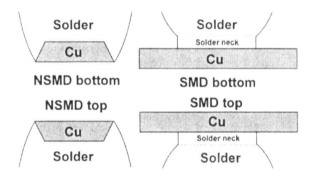

Fig. 4.7 Definitions of SMD and NSMD at the top and bottom of the joint.

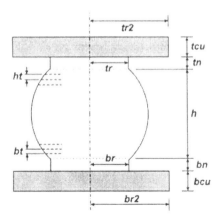

Fig. 4.8 Definitions of geometrical parameters in the SJGEO1 command for the SMD/SMD joint type (***jtyp*** = 1).

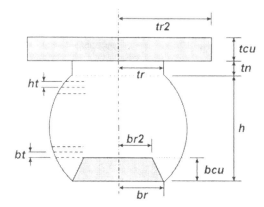

Fig. 4.9   Definitions of geometrical parameters in the SJGEO1 command for
the SMD/NSMD joint type (*jtyp* = 2).

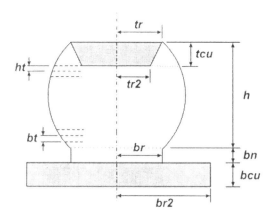

Fig. 4.10   Definitions of geometrical parameters in the SJGEO1 command for
the NSMD/SMD joint type (*jtyp* = 3).

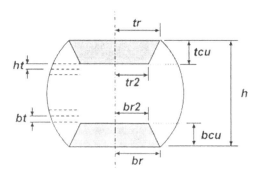

Fig. 4.11   Definitions of geometrical parameters in the SJGEO1 command for
the NSMD/NSMD joint type (*jtyp* = 4).

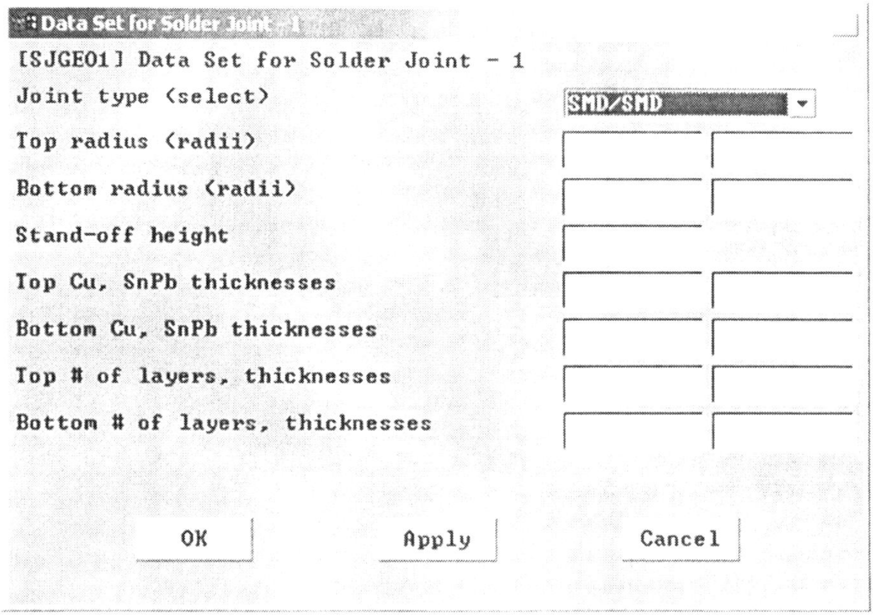

Fig. 4.12 Dialog box for SJGEO1.

SJGEO2, npt, r1, h1, r2, h2, r3, h3, ... , r7, h7, r8, h8, r9, h9

This is the second of three data sets describing the geometry and meshing preferences of the solder joints. The curved geometry of the solder joint is described using this command.

npt	number of input points
r1	radius of 1st point
h1	height of 1st point
r2	radius of 2nd point
h2	height of 2nd point
r3	radius of 3rd point
h3	height of 3rd point
r4	radius of 4th point
h4	height of 4th point
r5	radius of 5th point
h5	height of 5th point
r6	radius of 6th point
h6	height of 6th point
r7	radius of 7th point
h7	height of 7th point
r8	radius of 8th point
h8	height of 8th point
r9	radius of 9th point
h9	height of 9th point

The menu path for this command is as follows:

```
Preprocessor > Packaging > Solder Joint > -Data- Data Set 2
```

Notes: Radius and height information is required for the **npt** sample points excluding the end points of the curved boundary. The points should be given in order, starting from bottom to top. The definitions of the radii and heights are given for four different types of joints in Figs. 4.13-4.16. Figure 4.17 shows the dialog box for this command.

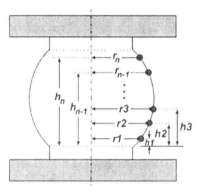

Fig. 4.13   Definitions of radii and heights of the points defining the curved joint surface in the SJGEO2 command for the SMD/SMD joint type (*jtyp* = 1).

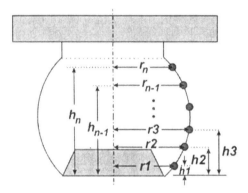

Fig. 4.14   Definitions of radii and heights of the points defining the curved joint surface in the SJGEO2 command for the SMD/NSMD joint type (*jtyp* = 2).

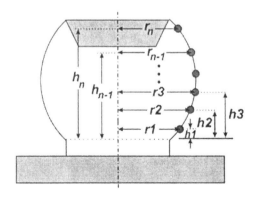

Fig. 4.15   Definitions of radii and heights of the points defining the curved joint surface in the SJGEO2 command for the NSMD/SMD joint type (*jtyp* = 3).

rn
rn-1
r3
r2
r1
hn
hn-1
h1
h2
h3

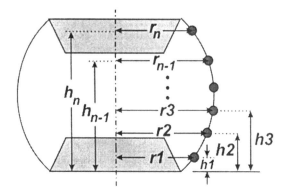

Fig. 4.16 Definitions of radii and heights of the points defining the curved joint surface in the SJGEO2 command for the NSMD/NSMD joint type (***jtyp*** = 4).

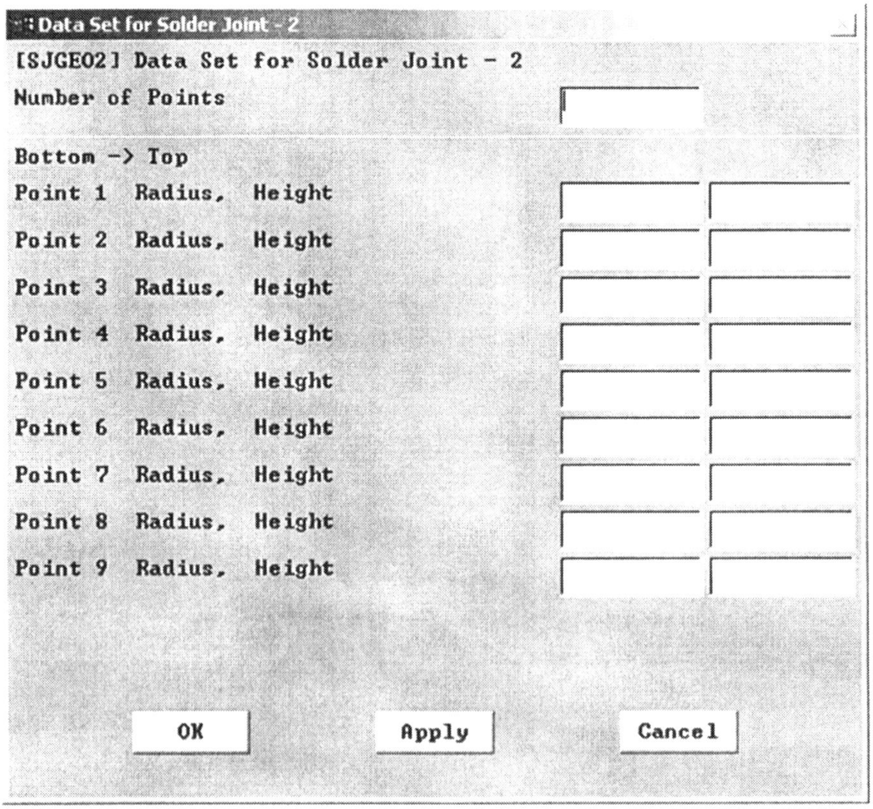

Data Set for Solder Joint - 2
[SJGEO2] Data Set for Solder Joint - 2
Number of Points
Bottom -> Top
Point 1 Radius, Height
Point 2 Radius, Height
Point 3 Radius, Height
Point 4 Radius, Height
Point 5 Radius, Height
Point 6 Radius, Height
Point 7 Radius, Height
Point 8 Radius, Height
Point 9 Radius, Height
OK
Apply
Cancel

Fig. 4.17 Dialog box for SJGEO2.

SJGEO3, *dr, drt, drb, dycut, dycub, dynt, dynb, dybt, dybb, dt, op1, dop1, op2, dop2*

This is the third of three data sets describing the geometry and meshing preferences of the solder joints. The number of divisions on the lines defining the solder geometry is specified using this command. Schematic descriptions of the number of divisions are given in Figs. 4.18-4.21 for all four joint types considered. The dialog box for SJGEO3 is shown in Fig. 4.22.

dr	number of divisions in the radial direction
drt	number of divisions in the radial direction for the portion of the top copper pad extending outside the solder area
drb	number of divisions in the radial direction for the portion of the bottom copper pad extending outside the solder area
dycut	number of divisions in the *y*-direction (thickness direction) at the top copper pad
dycub	number of divisions in the *y*-direction (thickness direction) at the bottom copper pad
dynt	number of divisions in the *y*-direction at the top solder neck layer
dynb	number of divisions in the *y*-direction at the bottom solder neck layer
dybt	number of divisions in the *y*-direction at the solder body
dybb	spacing ratio in the *y*-direction at the solder body (use of the spacing ratio is the same as in ANSYS®)
dt	number of divisions in the circumferential direction (must be an integer power of 2 but not less than 4)

op1	bottom mesh option—valid only for NSMD—if selected, the fillet area around the copper pad is meshed separately
	    0: do not use mesh option
	    1: use mesh option

dop1	number of divisions along the horizontal portion of the triangular boundary of the fillet area (refer to Figs. 4.19 and 4.21)

op2	top mesh option—valid only for NSMD—if selected, the fillet area around the copper pad is meshed separately
	    0: do not use mesh option
	    1: use mesh option

dop2	number of divisions along the horizontal portion of the triangular boundary of the fillet area (refer to Figs. 4.20 and 4.21)

The menu path for this command is as follows:

```
Preprocessor > Packaging > Solder Joint > -Data- Data Set 3
```

Notes: The number of divisions in the circumferential direction (**dt**) refers to the entire solder ball (complete circle). For example, if **dt** = 4 is specified, the solder ball mesh has 4 quarter circles as viewed from the top; if **dt** = 8 is specified, it has 8 octant circles; etc.

When the geometry of the solder balls possesses NSMD land, if the copper pad thickness is relatively large, the meshing of the solder may become impossible with the number of divisions specified in this command. In such cases, the user is advised to use the "mesh option." The *mesh option* creates a separate area for the fillet region around the copper pad, thus making the meshing possible. Figures 4.19-4.21 demonstrate the use of this option for different joint types.

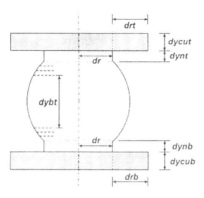

Fig. 4.18   Definitions of number of divisions on lines for the SJGEO3
command for the SMD/SMD joint type (*jtyp* = 1).

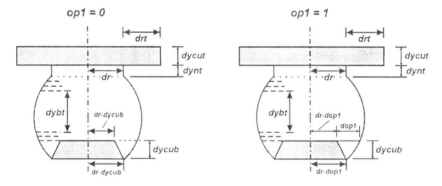

Fig. 4.19   Definitions of number of divisions on lines for the SJGEO3
command for the SMD/NSMD joint type (*jtyp* = 2).

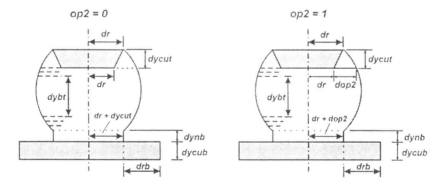

Fig. 4.20   Definitions of number of divisions on lines for the SJGEO3
command for the NSMD/SMD joint type (*jtyp* = 3).

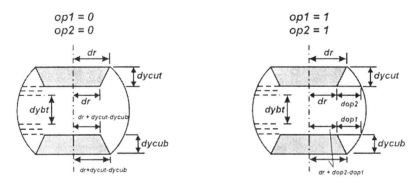

Fig. 4.21   Definitions of number of divisions on lines for the SJGEO3
command for the NSMD/NSMD joint type (*jtyp* = 4).

```
: Data Set for Solder Joint - 3

[SJGEO3] Data Set for Solder Joint - 3
Number of divisions
Radial                                          [        ]

Top Cu Radial #  2                              [        ]

Bottom Cu Radial #  2                           [        ]

Top Cu pad - in y                               [        ]

Bottom Cu pad - in y                            [        ]

Top SnPb layer - in y                           [        ]

Bottom SnPb layer - in y                        [        ]

SnPb body - in y                                [        ]

SnPb body - spacing ratio                       [        ]

Circumferential                                 [        ]

Bottom Meshing Option
                                            [ Cut Meshing
Bottom number of divisions                      [        ]

Top Meshing Option
                                            [ Cut Meshing
Top number of divisions                         [        ]



        OK                Apply              Cancel
```

Fig. 4.22   Dialog box for SJGEO3.

SJFULL, *x0, y0,z0, p, nx, nz, matsn, matcu, sym*

This macro (command) creates an array of solder joints based on the information obtained by the execution of SJGEO1, SJGEO2, and SJGEO3. Figures 4.23-4.24 provide different views of an array (2×2) of solder joints created by this command. The parameter *p* (pitch) is described in Fig. 4.23. Figure 4.25 shows the dialog box for this command.

x0	*x*-coordinate of the bottom center of the first solder joint
y0	*y*-coordinate of the bottom center of the first solder joint
z0	*z*-coordinate of the bottom center of the first solder joint
p	pitch
nx	number of solder joints in the *x*-direction
nz	number of solder joints in the *z*-direction
matsn	material number for solder
matcu	material number for copper pad
sym	symmetry type
	0: octant symmetry
	1: quarter symmetry
	2: half symmetry
	3: no symmetry

The menu path for this command is as follows:

```
Preprocessor > Packaging > Solder Joint > -Mesh- Solder Joint
```

Notes: The number of joints in the *x*- and *z*-directions does not refer to the actual package, instead, it is the number of solder joints to be modeled in ANSYS®. For example, if the actual package has a full array of 9 (in *x*) by 8 (in *z*) solder joints, and the rest of the package possesses quarter symmetry, then the user should choose quarter symmetry and specify *nx* = 5 and *nz* = 4, with *x0* = 0 and *z0* = *p*/2. At the end of the model generation, 4 half solder joints appear along the *y-z* plane. As a general rule, if the actual package has an even number of joints in one direction, then half that number should be specified in that direction. On the other hand, if the actual package has an odd number of joints in one direction, then the next integer higher than half that number should be specified in that direction.

The example shown in Figs. 4.23-4.24 possesses quarter symmetry. The planes of symmetry in ReliANS for different symmetry types are as follows:

Octant symmetry: *x-y* plane and ($x = z$)-plane.
Quarter symmetry: *x-y* and *y-z* planes
Half symmetry: *y-z* plane

In ReliANS, if the octant or quarter symmetry is selected, the model will be in the positive x-z-space and the symmetry planes will coincide with the global coordinate system planes. When half symmetry is used, the model will be in the positive x-space.

The *first solder joint* is the one that possesses the lowest x and z coordinates. When an array of solder joints is created, this solder joint will be meshed first. Then this mesh is copied in the positive x- and z-directions in increments of the pitch (p).

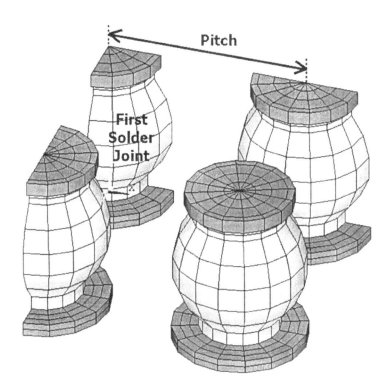

Fig. 4.23 Top oblique view of a 2×2 array of solder joints with quarter symmetry.

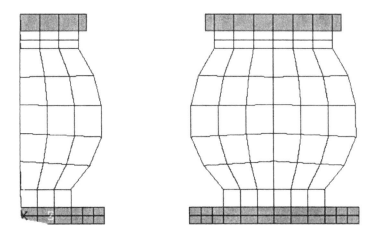

Fig. 4.24   Side (left) view of a 2 × 2 array of solder joints with quarter symmetry.

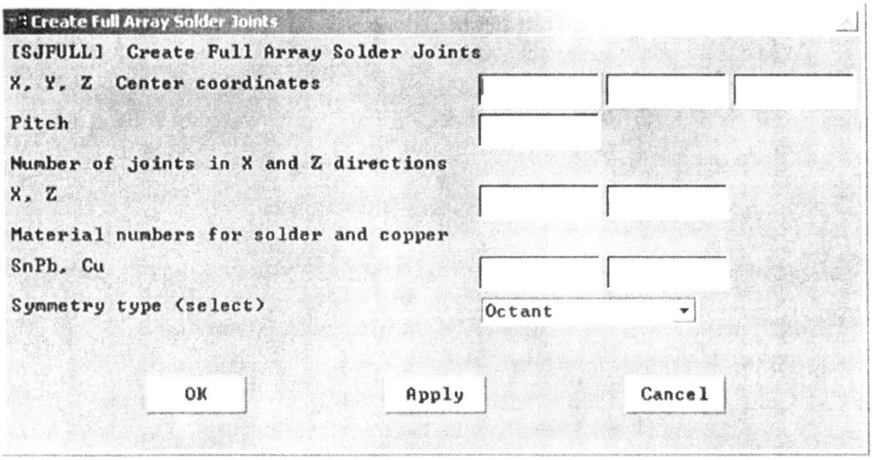

Fig. 4.25   Dialog box for SJFULL.

4.2.2   Motherboard Submenu Commands

FR4GEO1, *n, t1, t2, t3, ... , t15, t16, t17*

This is the first of three sets describing the geometry and meshing preferences of the printed circuit board (PCB) lay-up. This command specifies the
thickness values of an arbitrary number of layers (up to 17) of materials. A
schematic of the parameters is given in Fig. 4.26. Figure 4.27 shows the
dialog box for this command.

n	number of layers
t1	1st layer thickness
t2	2nd layer thickness
t3	3rd layer thickness
t4	4th layer thickness
t5	5th layer thickness
t6	6th layer thickness
t7	7th layer thickness
t8	8th layer thickness
t9	9th layer thickness
t10	10th layer thickness
t11	11th layer thickness
t12	12th layer thickness
t13	13th layer thickness
t14	14th layer thickness
t15	15th layer thickness
t16	16th layer thickness
t17	17th layer thickness

The menu path for this command is as follows:

```
Preprocessor > Packaging > Motherboard > -Data- Layer Data-1
```

Notes: The number of layers is arbitrary, with a maximum of 17. The
sequence is from bottom to top.

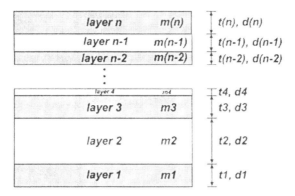

Fig. 4.26   Schematic of a PCB with layers.

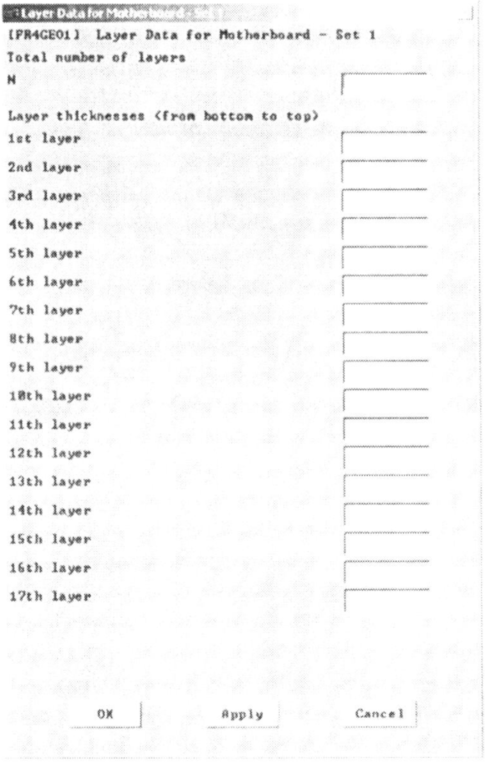

Fig. 4.27   Dialog box for FR4GEO1.

FR4GEO2, *n, d1, d2, d3, ... , d15, d16,d17*

This is the second of three data sets describing the geometry and meshing preferences of the printed circuit board (PCB) lay-up. This command specifies the number of divisions in the *y*-direction for the layers in the PCB. A schematic of the parameters is given in Fig. 4.26. Figure 4.28 shows the dialog box for this command.

n	number of layers
d1	1st layer number of divisions in the *y*-direction
d2	2nd layer number of divisions in the *y*-direction
d3	3rd layer number of divisions in the *y*-direction
d4	4th layer number of divisions in the *y*-direction
d5	5th layer number of divisions in the *y*-direction
d6	6th layer number of divisions in the *y*-direction
d7	7th layer number of divisions in the *y*-direction
d8	8th layer number of divisions in the *y*-direction
d9	9th layer number of divisions in the *y*-direction
d10	10th layer number of divisions in the *y*-direction
d11	11th layer number of divisions in the *y*-direction
d12	12th layer number of divisions in the *y*-direction
d13	13th layer number of divisions in the *y*-direction
d14	14th layer number of divisions in the *y*-direction
d15	15th layer number of divisions in the *y*-direction
d16	16th layer number of divisions in the *y*-direction
d17	17th layer number of divisions in the *y*-direction

The menu path for this command is as follows:

```
Preprocessor > Packaging > Motherboard > -Data- Layer Data-2
```

Notes: The sequence is from bottom to top. See BLCK7 for a schematic.

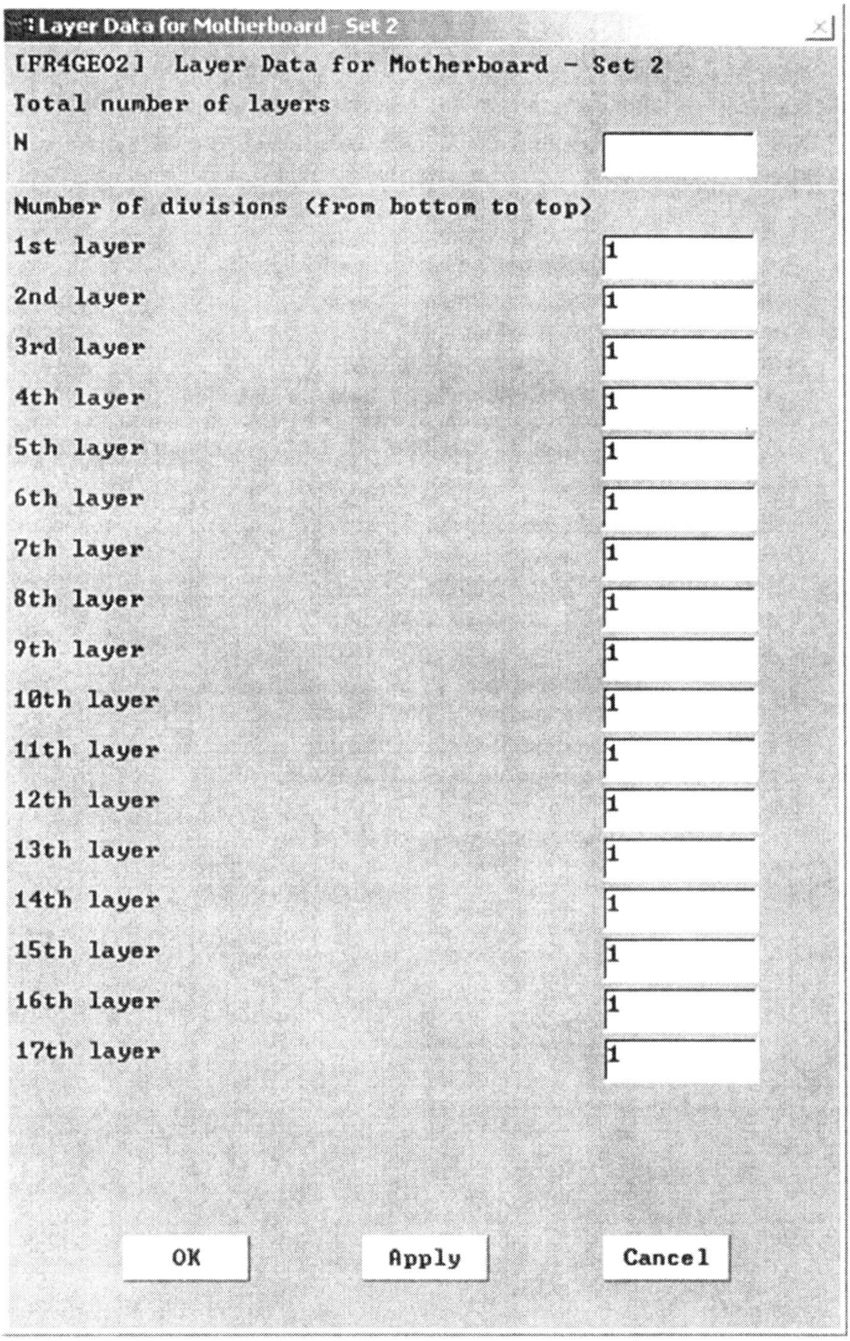

Fig. 4.28 Dialog box for FR4GEO2.

FR4GEO3, *n, m1, m2, m3, ... , m15, m16, m17*

This is the third of three data sets describing the geometry and meshing preferences of the printed circuit board (PCB) lay-up. This command specifies the material number for each of the layers in the PCB. A schematic of the parameters is given in Fig. 4.26. Figure 4.29 shows the dialog box for this command.

n	number of layers
m1	material number for 1st layer
m2	material number for 2nd layer
m3	material number for 3rd layer
m4	material number for 4th layer
m5	material number for 5th layer
m6	material number for 6th layer
m7	material number for 7th layer
m8	material number for 8th layer
m9	material number for 9th layer
m10	material number for 10th layer
m11	material number for 11th layer
m12	material number for 12th layer
m13	material number for 13th layer
m14	material number for 14th layer
m15	material number for 15th layer
m16	material number for 16th layer
m17	material number for 17th layer

The menu path for this command is as follows:

Preprocessor > Packaging > Motherboard > -Data- Layer Data-3

Notes: The sequence is from bottom to top. See BLCK7 for a schematic.

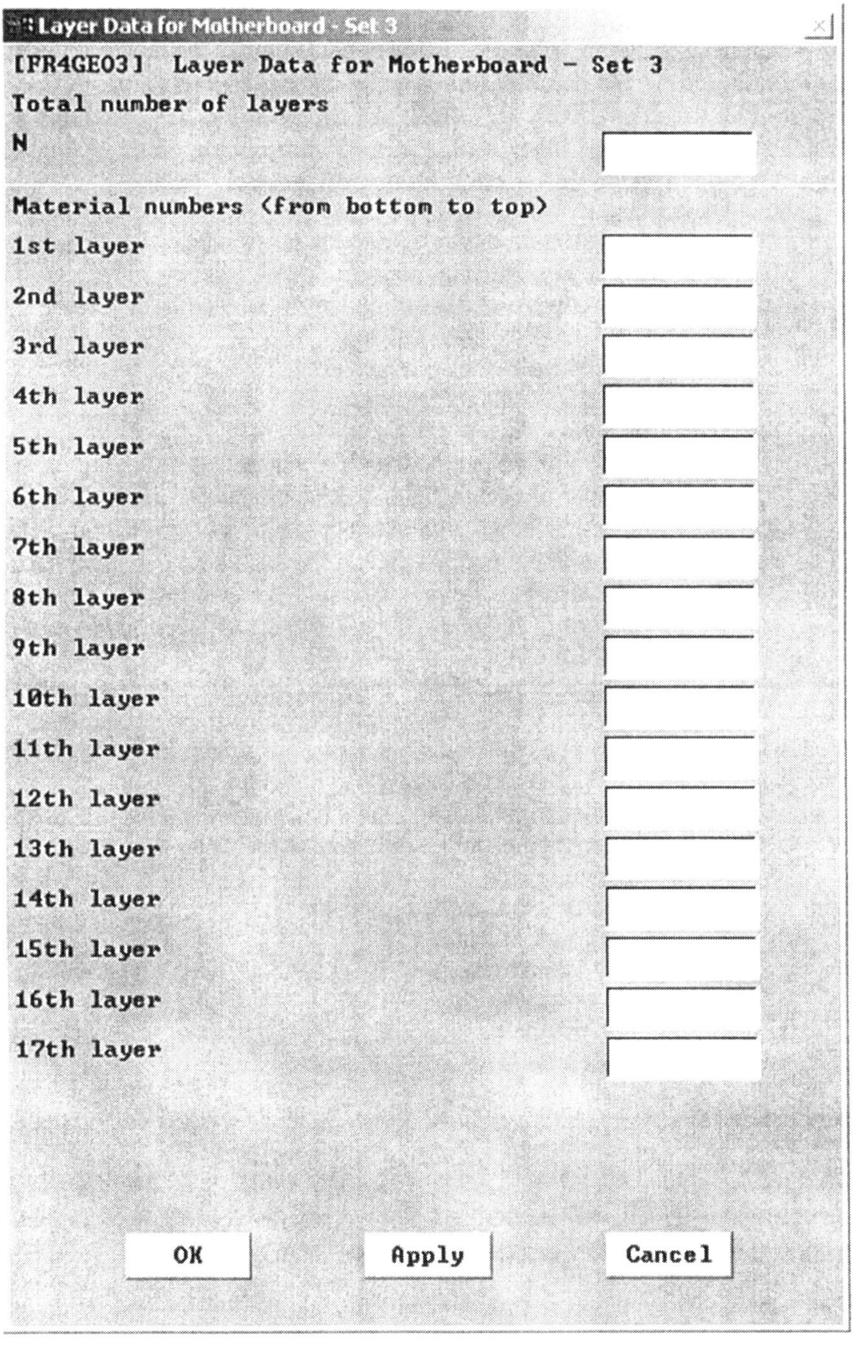

Fig. 4.29 Dialog box for FR4GEO3.

***BLCK7**, x0, y0, z0, p, nx, nz, rn, rcu, dr, dcu, dcu2, dt, sym*

This macro (command) creates an array of PCB blocks that have solder joint patterns based on the information obtained by the execution of FR4GEO1, FR4GEO2, and FR4GEO3. Schematics of the geometrical parameters for a 2×2 array of the PCB block with quarter symmetry are given in Fig. 4.30. Figure 4.31 shows the dialog box for this command.

x0	x-coordinate of the bottom center of the first block with a solder joint mesh pattern
y0	y-coordinate of the bottom center of the first block with a solder joint mesh pattern
z0	z-coordinate of the bottom center of the first block with a solder joint mesh pattern
p	pitch
nx	number of solder joints in the x-direction
nz	number of solder joints in the z-direction
rn	radius of the circular solder area where it meets the PCB
rcu	radius of the copper pad on the PCB side (if different from the solder)
dr	number of radial divisions in the solder area where it meets the PCB
dcu	number of divisions in the radial direction for the portion of the copper pad area extending outside the solder area
dcu2	number of divisions in the radial direction for the portion of the PCB outside the copper pad area
dt	number of divisions in the circumferential direction (must be an integer power of 2 but not less than 4)
sym	symmetry type 0: octant symmetry 1: quarter symmetry 2: half symmetry 3: no symmetry

The menu path for this command is as follows:

```
Preprocessor > Packaging > Motherboard > -Mesh- Full Array
```

Notes: In Fig. 4.30, ***dr*** = 3, ***dcu*** = 2, and ***dcu2*** = 2. The discussions covered in SJFULL on the number of joints in the x- and z-directions, the array, and the symmetry are valid for this command, also.

pitch
rn,dr
rcu,dcu
dcu2
Material 4
Material 3
Material 2
Material 1
d4=3
d3=2
d2=1
d1=2

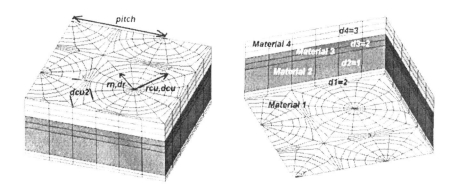

Fig. 4.30 Top (left) and bottom (right) oblique view of a 2 × 2 array of a PCB with a solder mesh pattern.

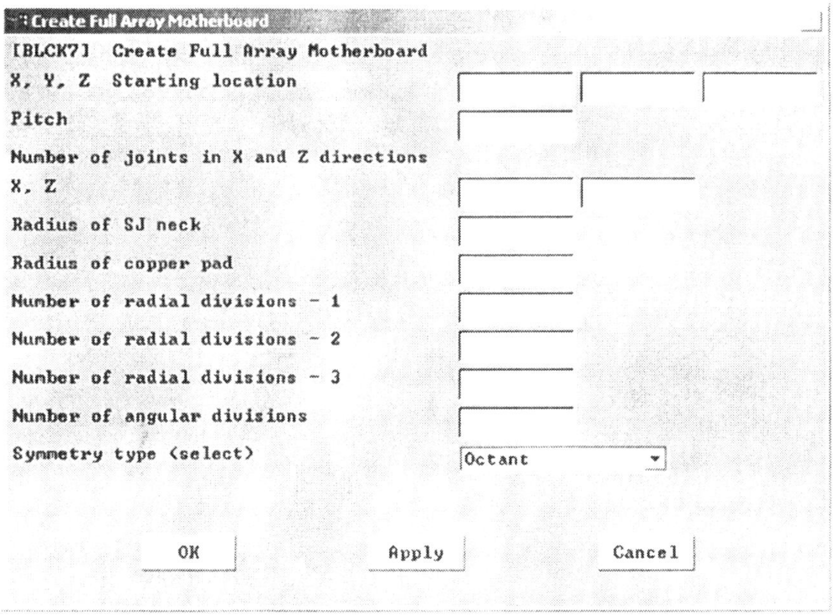

Create Full Array Motherboard
[BLCK7] Create Full Array Motherboard
X, Y, Z Starting location
Pitch
Number of joints in X and Z directions
X, Z
Radius of SJ neck
Radius of copper pad
Number of radial divisions - 1
Number of radial divisions - 2
Number of radial divisions - 3
Number of angular divisions
Symmetry type (select)
Octant
OK
Apply
Cancel

Fig. 4.31 Dialog box for BLCK7.

BLCK8, x0, y0, z0, lx, lz, dx, dz

This macro (command) creates a rectangular PCB block based on the information (thickness and number of divisions in the thickness direction) obtained by execution of FR4GEO1, FR4GEO2, and FR4GEO3. Figure 4.32 shows the dialog box for this command.

x0	*x*-coordinate of the bottom corner of the block
y0	*y*-coordinate of the bottom corner of the block
z0	*z*-coordinate of the bottom corner of the block
lx	length in the *x*-direction
lz	length in the *z*-direction
dx	number of divisions in the *x*-direction
dz	number of divisions in the *z*-direction

The menu path for this command is as follows:

```
Preprocessor > Packaging > Motherboard > -Mesh- Rectangular
```

Notes: The starting point of the rectangular block created by BLCK8 refers to the corner that has the lowest *x,y,z* coordinates. The build-up is done from bottom to top. Figure 4.33 shows an example mesh created by this command using $x0 = 1$, $y0 = 2$, $z0 = 3$, $lx = 20$, $lz = 40$, $dx = 4$, and $dz = 5$.

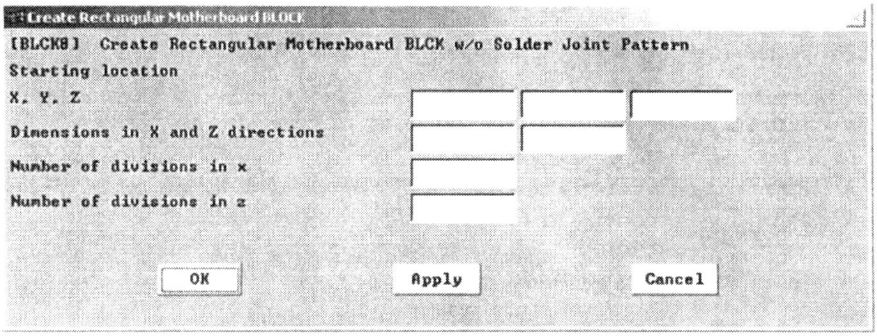

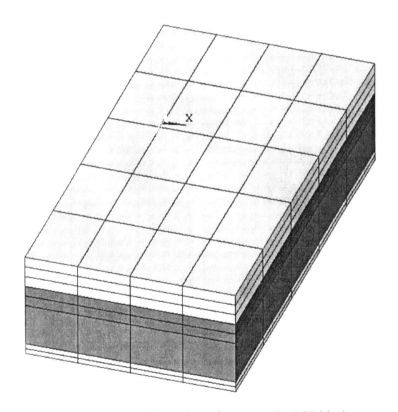

Fig. 4.32 Dialog box for BLCK8.

Fig. 4.33 Top oblique view of a rectangular PCB block.

BLCK9, *x0, y0, z0, lx, lz, theta, dx, dz*

This macro (command) creates a trapezoidal PCB block based on the information (thickness and number of divisions in the thickness direction) obtained by execution of FR4GEO1, FR4GEO2, and FR4GEO3. Figure 4.34 shows the dialog box for this command.

x0	*x*-coordinate of the bottom corner of the block
y0	*y*-coordinate of the bottom corner of the block
z0	*z*-coordinate of the bottom corner of the block
lx	length in the *x*-direction
lz	length in the *z*-direction
theta	angle of the trapezoid in degrees (refer to Fig. 4.35)
dx	number of divisions in the *x*-direction
dz	number of divisions in the *z*-direction

The menu path for this command is as follows:

```
Preprocessor > Packaging > Motherboard > -Mesh- Trapezoidal
```

Notes: The starting point of the trapezoidal block created by BLCK9 refers to the corner that has the lowest *x,y,z* coordinates. The build-up is achieved from bottom to top. Figure 4.35 shows an example mesh created by this command using ***x0*** = 1, ***y0*** = 2, ***z0*** = 3, ***lx*** = 20, ***lz*** = 40, ***theta*** = 60, ***dx*** = 4, and ***dz*** = 5.

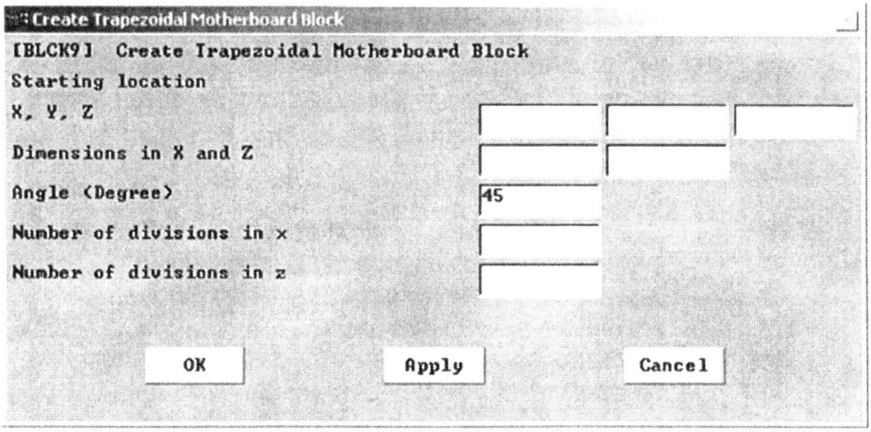

Fig. 4.34   Dialog box for BLCK9.

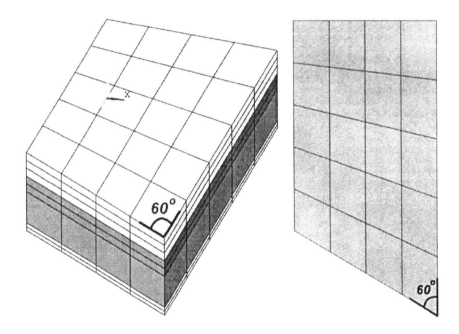

Fig. 4.35   Top oblique (left) and top views (right) of a trapezoidal
PCB block.

BLCK10, *x0, y0, z0, lxz, dxz*

This command creates a triangular PCB block based on the information (thickness and number of divisions in the thickness direction) obtained by the execution of FR4GEO1, FR4GEO2, and FR4GEO3. Figure 4.36 shows the dialog box for this command.

x0	*x*-coordinate of the bottom corner of the block
y0	*y*-coordinate of the bottom corner of the block
z0	*z*-coordinate of the bottom corner of the block
lxz	length in the *x*- and *z*-directions
dxz	number of divisions in the *x*- and *z*-directions (must be even)

The menu path for this command is as follows:

```
Preprocessor > Packaging > Motherboard > -Mesh- Triangular
```

Notes: The starting point of the triangular block created by BLCK10 refers to the corner that has the lowest x,y,z coordinates. Because mapped meshing is used, the user must specify an even number of divisions for the sides of the triangle. The build-up is achieved from bottom to top. Figure 4.37 shows an example mesh created by this command using **x0** = 1, **y0** = 2, **z0** = 3, **lxz** = 20, and **dxz** = 6.

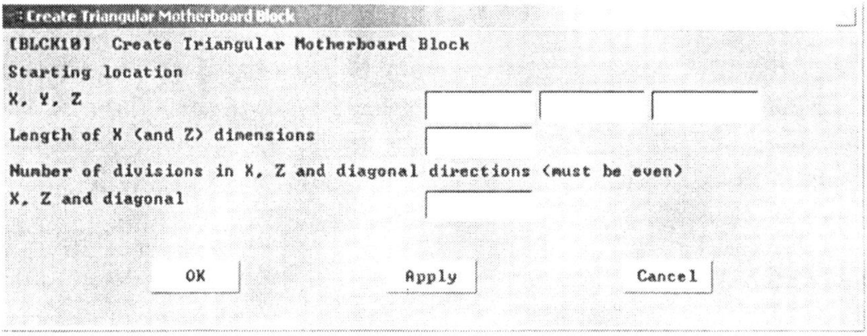

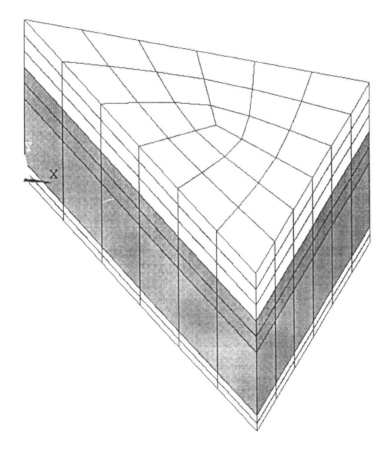

Fig. 4.36 Dialog box for BLCK10.

Fig. 4.37 Top oblique view of a triangular PCB block.

***BLCK11**, x1, z1, x2, z2, x3, z3, x4, z4, y0, d12_34, d23_41*

This command creates a general block of PCB based on the information (thickness and number of divisions in the thickness direction) obtained by the execution of FR4GEO1, FR4GEO2, and FR4GEO3. The geometry of a general block defined by its corner coordinates is described in Fig. 4.38. Figure 4.39 shows the dialog box for this command.

x1	$x1$-coordinate
z1	$z1$-coordinate
x2	$x2$-coordinate
z2	$z2$-coordinate
x3	$x3$-coordinate
z3	$z3$-coordinate
x4	$x4$-coordinate
z4	$z4$-coordinate
y0	y-coordinate of the bottom surface (usually zero)
d12_34	Number of divisions between lines [1-2] and [3-4]
d23_41	Number of divisions between lines [2-3] and [4-1]

The menu path for this command is as follows:

```
Preprocessor > Packaging > Motherboard > -Mesh- General Block
```

Notes: The starting point of the general PCB block created by BLCK11 is an arbitrary point on the bottom surface. The coordinates should be given in a clockwise or counterclockwise (not arbitrary) order. However, the starting point can be arbitrary. The build-up is achieved from bottom to top. Figure 4.38 shows an example mesh created by this command using $x1 = 10$, $z1 = 6$, $x2 = 1$, $z2 = 1$, $x3 = 2$, $z3 = 15$, $x4 = 12$, $z4 = 10$, $y0 = 1$, $d12\_34 = 4$, and $d23\_41 = 6$.

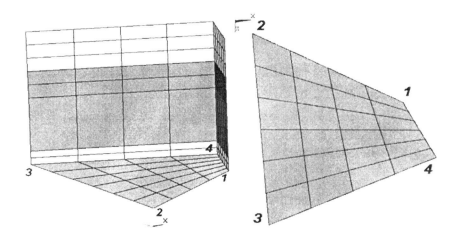

Fig. 4.38 Bottom oblique (left) and top (right) views of a general PCB block.

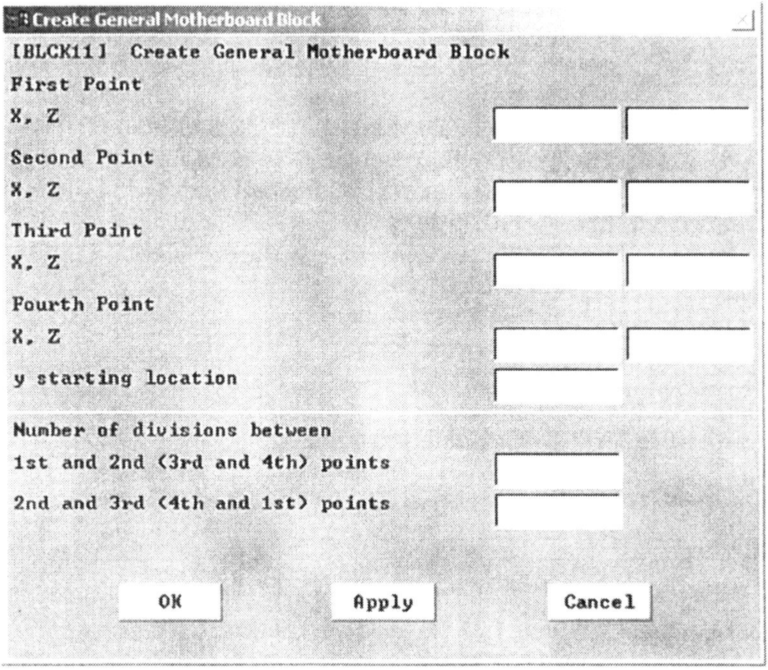

Fig. 4.39 Dialog box for BLCK 11.

4.2.3   Layer Submenu Commands

BLCK1, *x0, y0, z0, p, nx, nz, rn, rcu, t, dsn, dcu, dcu2, dy, dt, mat, sym*

This command creates a full array of blocks that have a solder joint pattern. Figure 4.40 shows the dialog box for this command.

x0	x-coordinate of the bottom center of the first block with a solder joint mesh pattern
y0	y-coordinate of the bottom center of the first block with a solder joint mesh pattern
z0	z-coordinate of the bottom center of the first block with a solder joint mesh pattern
p	pitch
nx	number of solder joints in the x-direction
nz	number of solder joints in the z-direction
rn	radius of the corresponding circular solder area
rcu	radius of the corresponding copper pad (if different from the solder radius)
t	thickness of the layer (in the y-direction)
dsn	number of radial divisions in the corresponding solder area
dcu	number of divisions in the radial direction for the portion of the copper pad area extending outside the solder area
dcu2	number of divisions in the radial direction for the portion outside the copper pad area
dy	number of divisions in the thickness direction (y-direction)
dt	number of divisions in the circumferential direction (must be an integer power of 2 but not less than 4)
mat	material number for the layer
sym	symmetry type
	0: octant symmetry
	1: quarter symmetry
	2: half symmetry
	3: no symmetry

The menu path for this command is as follows:

```
Preprocessor > Packaging > Layer > Create Layer
```

Notes: The explanations in SJFULL on the number of joints in the x- and z-directions, the array of solder balls, and the symmetry conditions are valid for this command. Figure 4.41 shows an example mesh created by this command using **x0** = 0, **y0** = 2, **z0** = 0, **p** = 20, **nx** = 2, **nz** = 2, **rn** = 5, **rcu** = 7, **t** = 2, **dsn** = 4, **dcu** = 3, **dcu2** = 2, **dy** = 2, **dt** = 16, **mat** = 3, and **sym** = 1.

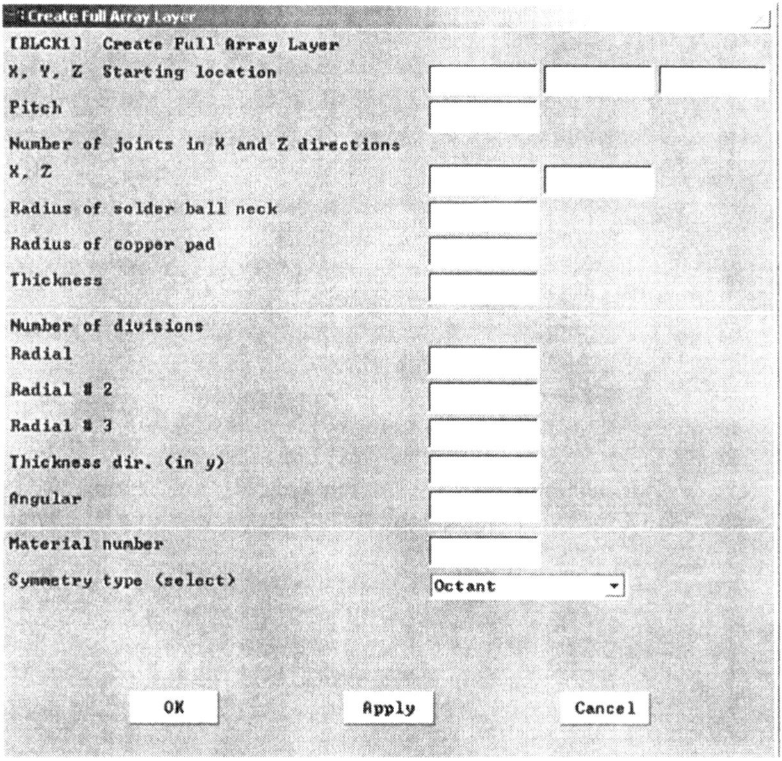

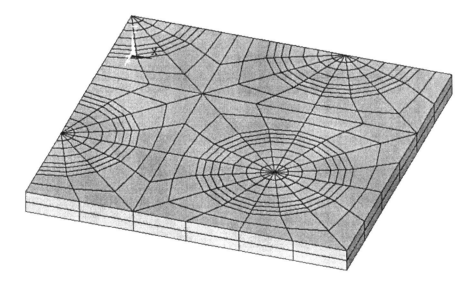

Fig. 4.40   Dialog box for BLCK1.

Fig. 4.41   Top oblique view of a 2 × 2 array of a layer with a solder mesh pattern.

BLCK2, *x0, y0, z0, p, nx, nz, rn, rcu, tcu, phi, dcu, dcu2, dy, dt, mat, sym*

This command creates a full array of blocks that have a solder joint pattern without the circular center portion. This command is designed specifically to model a solder mask. Figure 4.42 shows the dialog box for this command.

x0	x-coordinate of the bottom center of the first block with a solder joint mesh pattern
y0	y-coordinate of the bottom center of the first block with a solder joint mesh pattern
z0	z-coordinate of the bottom center of the first block with a solder joint mesh pattern
p	pitch
nx	number of solder joints in the x-direction
nz	number of solder joints in the z-direction
rn	radius of the corresponding circular solder area
rcu	radius of the corresponding copper pad (if different from the solder radius)
tcu	thickness of the layer (in the y-direction)
phi	angle of inclination in degrees (in order to simulate the detachment of the solder mask from the solder in the thermal cycle)—refer to Fig. 4.43.
dcu	number of divisions in the radial direction for the portion of copper pad area extending outside the solder area
dcu2	number of divisions in the radial direction for the portion of the PCB outside the copper pad area
dy	number of divisions in the thickness direction (y-direction)
dt	number of divisions in the circumferential direction (must be an integer power of 2 but not less than 4)
mat	material number for the layer
sym	symmetry type
	0: octant symmetry
	1: quarter symmetry
	2: half symmetry
	3: no symmetry

The menu path for this command is as follows:

```
Preprocessor > Packaging > Layer > Solder Mask
```

Notes: Figure 4.43 shows a 2 × 2 array of blocks with octant symmetry. The explanations in SJFULL on the number of joints in the x- and z-directions, the array of solder balls, and the symmetry conditions are valid for this command.

Angle *phi* is the inclination angle of the inner walls of the hollow center portion measured from the horizontal. When there is no inclination (vertical walls), this value becomes 90 degrees. Figure 4.43 provides clarification, and shows an example mesh created by this command using *x0* = 0, *y0* = 2, *z0* = 0, *p* = 20, *nx* = 2, *nz* = 2, *rn* = 5, *rcu* = 7, *tcu* = 2, *phi* = 60, *dcu* = 2, *dcu2* = 4, *dy* = 2, *dt* = 16, *mat* = 4, and *sym* = 0.

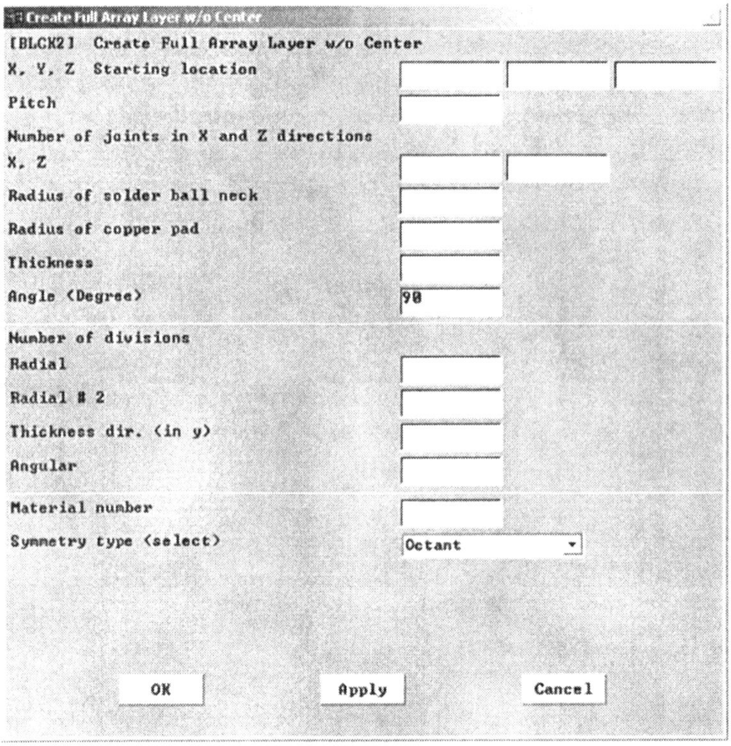

Create Full Array Layer w/o Center

[BLCK2] Create Full Array Layer w/o Center

X, Y, Z Starting location

Pitch

Number of joints in X and Z directions
X, Z

Radius of solder ball neck

Radius of copper pad

Thickness

Angle (Degree)   90

Number of divisions

Radial

Radial # 2

Thickness dir. (in y)

Angular

Material number

Symmetry type (select)   Octant

OK   Apply   Cancel

Fig. 4.42   Dialog box for BLCK2.

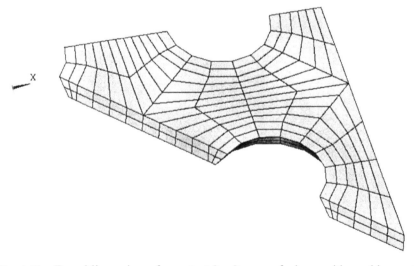

Fig. 4.43   Top oblique view of an octant 2 × 2 array of a layer with a solder mesh pattern without the center portion. Note the slanted (60°) inner walls of the mesh.

DISK1, *x0*, *y0*, *z0*, *p*, *nx*, *nz*, *ri*, *ro*, *t*, *di*, *do*, *dy*, *dt*, *mat*, *sym*

This command creates a full array of cylindrical blocks that have solder joint and copper pad mesh patterns. This command is designed specifically to model copper pads. Figure 4.44 shows the dialog box for this command.

x0	x-coordinate of the bottom center of the first block with a solder joint mesh pattern
y0	y-coordinate of the bottom center of the first block with a solder joint mesh pattern
z0	z-coordinate of the bottom center of the first block with a solder joint mesh pattern
p	pitch
nx	number of disks in the x-direction
nz	number of disks in the z-direction
ri	radius of the corresponding circular solder area
ro	radius of the corresponding copper pad
t	thickness of the disks (in the y-direction)
di	number of radial divisions for the inner disk
do	number of radial divisions for the portion between the inner disk and the edge of the outer disk
dy	number of divisions in the thickness direction (y-direction)
dt	number of divisions in the circumferential direction (must be an integer power of 2 but not less than 4)
mat	material number for the layer
sym	symmetry type
	0: octant symmetry
	1: quarter symmetry
	2: half symmetry
	3: no symmetry

The menu path for this command is as follows:

```
Preprocessor > Packaging > Layer > Disks
```

Notes: Figure 4.45 shows a 2 × 2 array of disks with quarter symmetry. The explanations in SJFULL on the number of joints in the *x*- and *z*-directions, the array of solder balls, and the symmetry conditions are valid for this command. The example mesh was created by this command using *x0* = 0, *y0* = 2, *z0* = 0, *p* = 20, *nx* = 2, *nz* = 2, *ri* = 5, *ro* = 7, *t* = 2, *di* = 2, *do* = 4, *dy* = 2, *dt* = 16, *mat* = 2, and *sym* = 1.

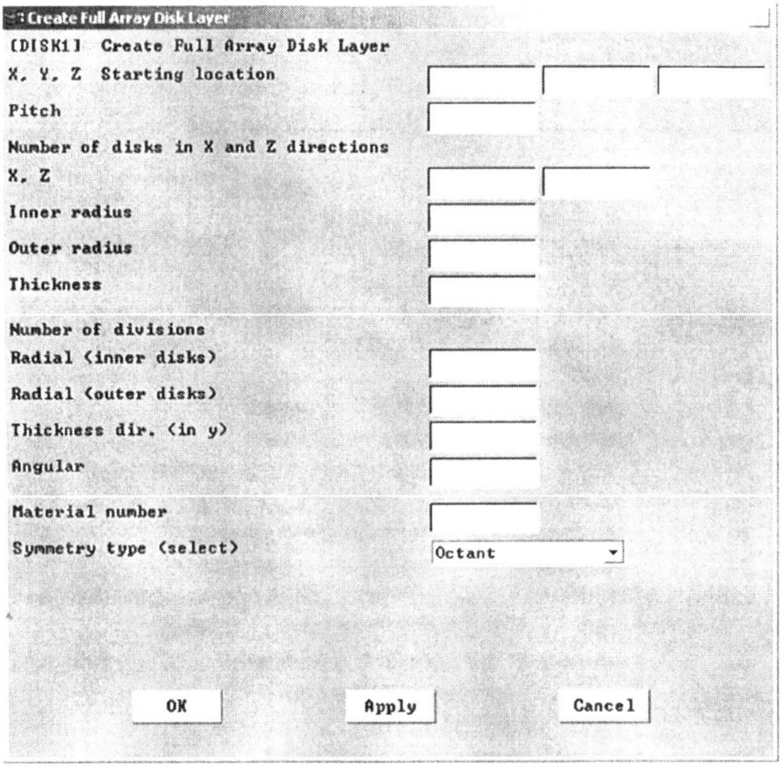

Create Full Array Disk Layer

[DISK1] Create Full Array Disk Layer

X, Y, Z Starting location

Pitch

Number of disks in X and Z directions

X, Z

Inner radius

Outer radius

Thickness

Number of divisions
Radial (inner disks)

Radial (outer disks)

Thickness dir. (in y)

Angular

Material number

Symmetry type (select)   Octant

        OK   Apply   Cancel

Fig. 4.44   Dialog box for DISK1.

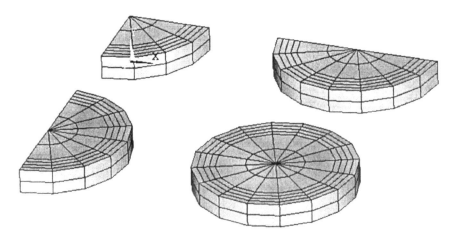

Fig. 4.45   Top oblique view of a quarter-symmetry 2 × 2 array of disks.

DISK2, x0, y0, z0, p, nx, nz, ri, ro, t, phi, do, dy, dt, mat, sym

This command creates a full array of cylindrical hollow blocks that have solder joint and copper pad mesh patterns. Figure 4.46 shows the dialog box for this command.

x0	x-coordinate of the bottom center of the first block with a solder joint mesh pattern
y0	y-coordinate of the bottom center of the first block with a solder joint mesh pattern
z0	z-coordinate of the bottom center of the first block with a solder joint mesh pattern
p	pitch
nx	number of disks in the x-direction
nz	number of disks in the z-direction
ri	radius of the corresponding circular solder area
ro	radius of the corresponding copper pad
t	thickness of the disks (in the y-direction)
phi	angle of inclination in degrees (in order to simulate detachment of the solder mask from the solder in the thermal cycle—refer to Fig. 4.47
do	number of radial divisions for the portion between the inner disk and the edge of the outer disk
dy	number divisions in the thickness direction (y-direction)
dt	number of divisions in the circumferential direction (must be an integer power of 2 but not less than 4)
mat	material number for the layer
sym	symmetry type
	0: octant symmetry
	1: quarter symmetry
	2: half symmetry
	3: no symmetry

The menu path for this command is as follows:

```
Preprocessor > Packaging > Layer > Hollow Disks
```

Notes: Figure 4.47 shows a 2×2 array of disks with quarter symmetry. The explanations in SJFULL on the number of joints in the x- and z-directions, the array of solder balls, and the symmetry conditions are valid for this command. The example mesh was created by this command using **x0** = 0, **y0** = 2, **z0** = 0, **p** = 20, **nx** = 2, **nz** = 2, **ri** = 5, **ro** = 7, **t** = 2, **phi** = 60, **do** = 4, **dy** = 2, **dt** = 16, **mat** = 5, and **sym** = 1.

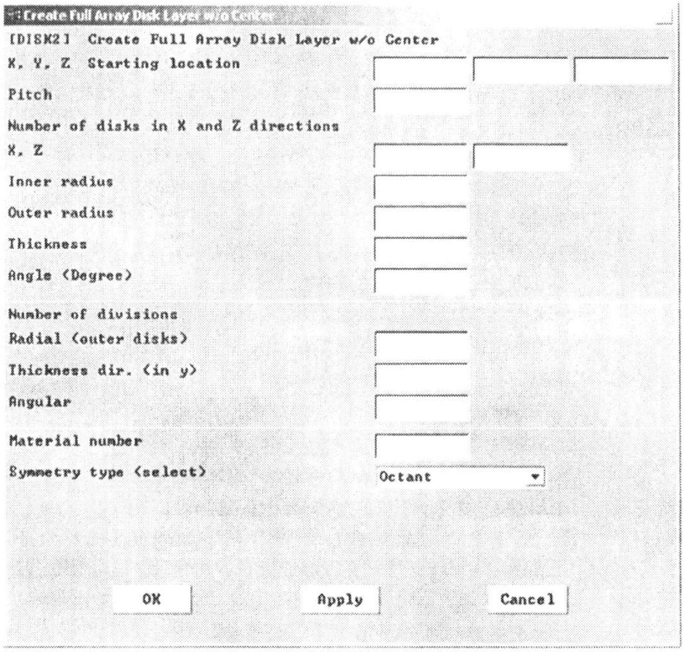

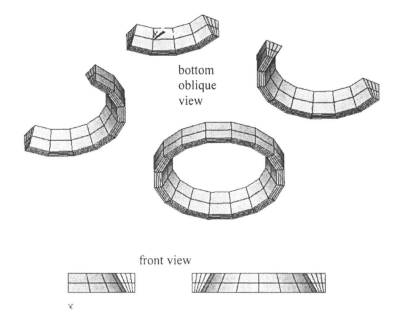

Fig. 4.46   Dialog box for DISK2.

bottom
oblique
view

front view

Fig. 4.47   Bottom oblique and front views of a quarter-symmetry 2×2 array of
hollow disks with a solder mesh pattern without the center portion.
Note the slanted (60°) inner walls of the mesh.

BLCK3, x0, y0, z0, lx, ly, lz, dx, dy, dz, mat

This command creates a rectangular layer. Figure 4.48 shows the dialog box for this command.

x0	x-coordinate of the bottom corner of the block
y0	y-coordinate of the bottom corner of the block
z0	z-coordinate of the bottom corner of the block
lx	length in the x-direction
ly	length in the y-direction
lz	length in the z-direction
dx	number of divisions in the x-direction
dy	number of divisions in the y-direction
dz	number of divisions in the z-direction
mat	material number for the layer

The menu path for this command is as follows:

```
Preprocessor > Packaging > Layer > Rectangular
```

Notes: Refer to BLCK8 for examples because this command is similar to the BLCK8 command.

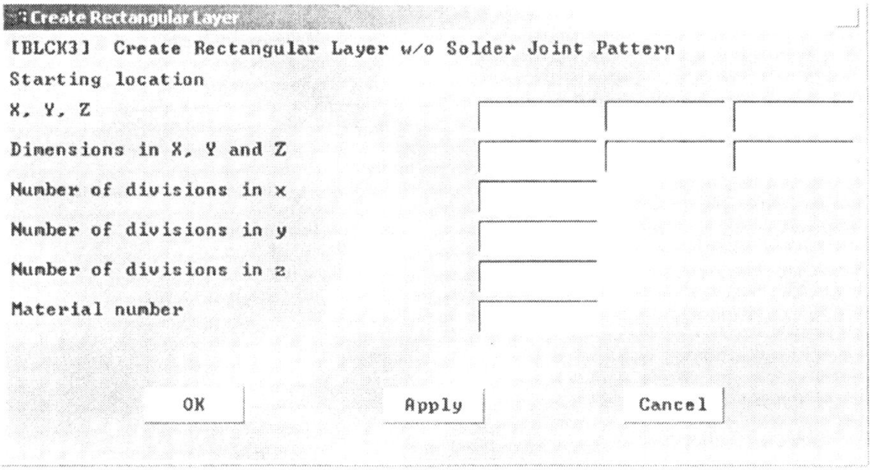

Fig. 4.48 Dialog box for BLCK3.

BLCK4, *x0, y0, z0, lx, ly, lz, theta, dx, dy, dz, mat*

This command creates a trapezoidal block layer. Figure 4.49 shows the dialog box for this command.

x0	*x*-coordinate of the bottom corner of the block
y0	*y*-coordinate of the bottom corner of the block
z0	*z*-coordinate of the bottom corner of the block
lx	length in the *x*-direction
ly	length in the *y*-direction
lz	length in the *z*-direction
theta	angle of the trapezoid in degrees (refer to Fig. 4.35)
dx	number of divisions in the *x*-direction
dy	number of divisions in the *y*-direction
dz	number of divisions in the *z*-direction
mat	material number

The menu path for this command is as follows:

```
Preprocessor > Packaging > Layer > Trapezoidal
```

Notes: Refer to BLCK9 for examples because this command is similar to the BLCK9 command.

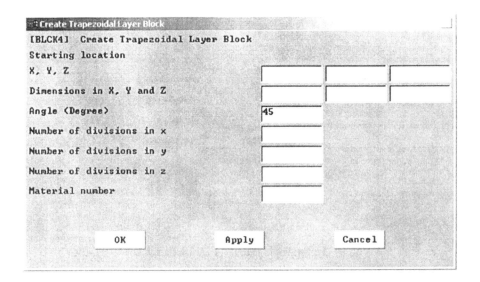

Fig. 4.49   Dialog box for BLCK4.

***BLCK5**, x0, y0, z0, lxz, ly, dxz, dy, mat*

This command creates a triangular layer. Figure 4.50 shows the dialog box for this command.

x0	x-coordinate of the bottom corner of the block
y0	y-coordinate of the bottom corner of the block
z0	z-coordinate of the bottom corner of the block
lxz	length in the x- and z-directions
ly	thickness of the layer
dxz	number of divisions in the x- and z-directions (must be even)
dy	number of divisions in the y-direction
mat	material number for the layer

The menu path for this command is as follows:

```
Preprocessor > Packaging > Layer > Triangular
```

Notes: Refer to BLCK10 for examples because this command is similar to the BLCK10 command.

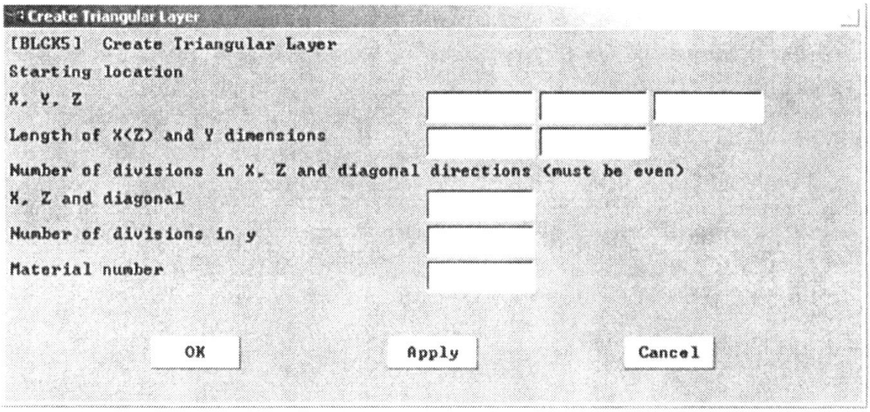

Fig. 4.50 Dialog box for BLCK5.

BLCK6, *x1, z1, x2, z2, x3, z3, x4, z4, y0, t, d12_34, d23_41, dy, mat*

This command creates a general block layer. The geometry of a general block defined by its corner coordinates is described in Fig. 4.38. Figure 4.51 shows the dialog box for this command.

x1	$x1$-coordinate
z1	$z1$-coordinate
x2	$x2$-coordinate
z2	$z2$-coordinate
x3	$x3$-coordinate
z3	$z3$-coordinate
x4	$x4$-coordinate
z4	$z4$-coordinate
y0	y-coordinate of the bottom surface of the layer
t	thickness of the layer
d12_34	number of divisions between lines [1-2] and [3-4]
d23_41	number of divisions between lines [2-3] and [4-1]
dy	number of divisions in the y-direction
mat	material number for the layer

The menu path for this command is as follows:

```
Preprocessor > Packaging > Layer > General Block
```

Notes: Refer to BLCK11 for examples because this command is similar to the BLCK11 command.

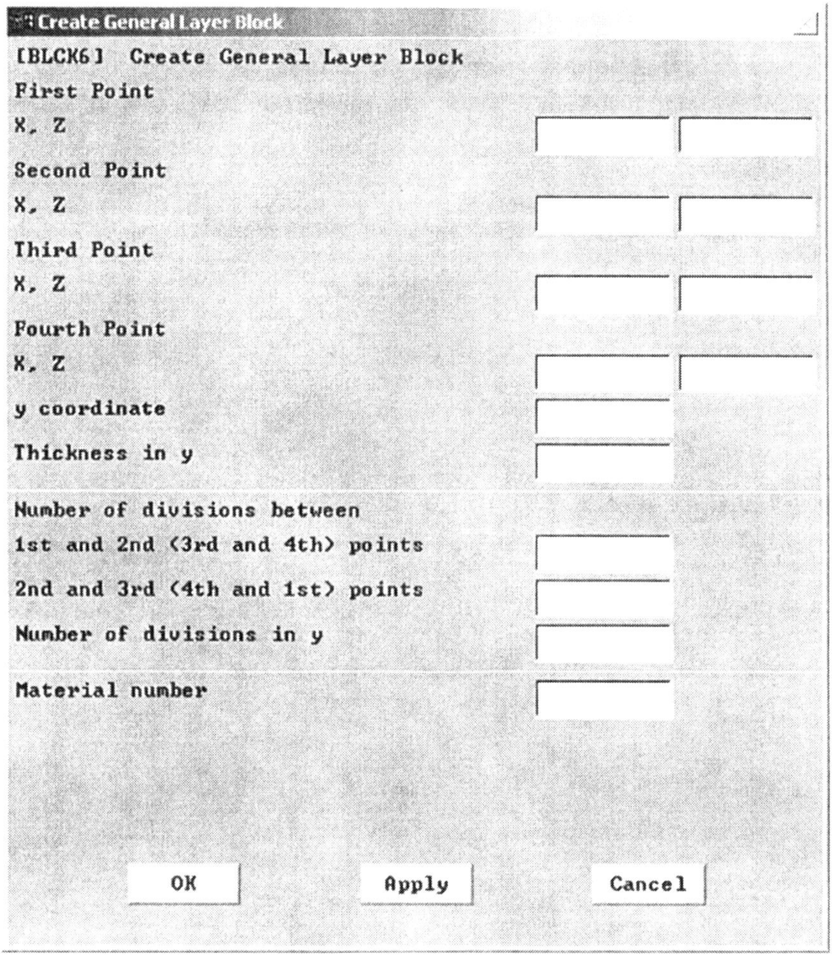

Fig. 4.51 Dialog box for BLCK6.

4.2.4 Material Properties Submenu Commands

EPMAT, *mat, label*

This command assigns material properties from the ReliANS database for commonly used materials. Figure 4.52 shows the dialog box for this command.

mat   material number
label   label (corresponding number in the list)

The menu path for this command is as follows:

```
Preprocessor > Packaging > Material Props > Material Props
```

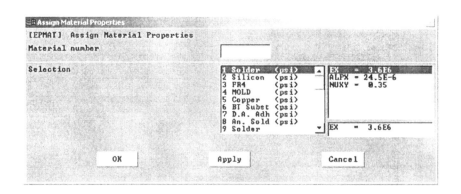

Fig. 4.52   Dialog box for EPMAT.

USMAT, *mat, label*

This command assigns material properties added to the ReliANS database by the user. Figure 4.53 shows the dialog box for this command.

mat	material number
label	label

The menu path for this command is as follows:

```
Preprocessor > Packaging > Material Props > User Mat Props
```

Notes: In order to use this command, the user must edit the "epmat.mac" file with a text editor and add material properties in the same format as the ones already there. It requires some knowledge of the ANSYS® Parametric Design Language (APDL). If the user chooses not to take this route, he/she should specify material properties using conventional ANSYS®-GUI methods/menus.

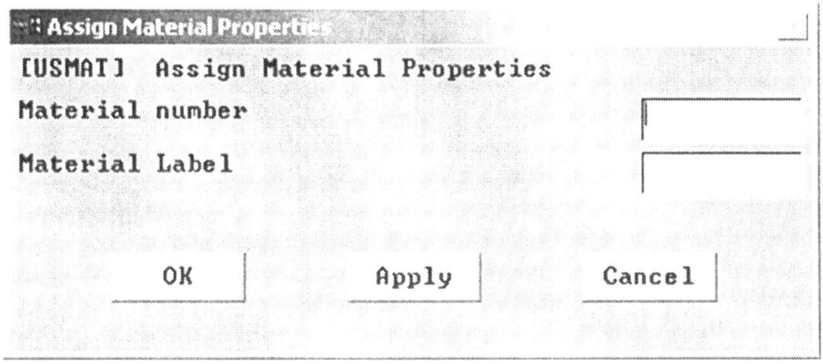

Fig. 4.53 Dialog box for USMAT.

4.2.5   Additional Menu Commands

EPEDEL

This command deletes the set of elements created by the last ReliANS command/menu item. It is designed to compensate for the lack of an "undo" function in the ANSYS® structure. Figure 4.54 shows the dialog box for this command.

The menu path for this command is as follows:

```
Preprocessor > Packaging > Delete elems
```

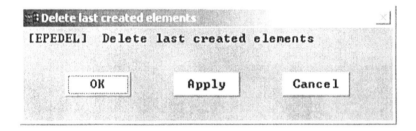

Fig. 4.54   Dialog box for EPEDEL.

CHGVISC, *mat*

This command changes the element type of the specified material from SOLID45 to VISCO107. It is designed specifically for the solder material. Figure 4.55 shows the dialog box for this command.

    mat   material number for nonlinear solder

The menu path for this command is as follows:

```
Preprocessor > Packaging > Change to Visco
```

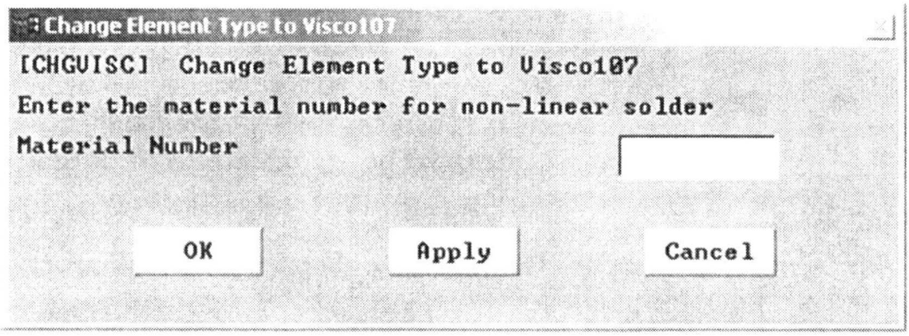

```
Change Element Type to Visco107
[CHGUISC]  Change Element Type to Uisco107
Enter the material number for non-linear solder
Material Number

              OK              Apply              Cancel
```

Fig. 4.55   Dialog box for CHGVISC.

4.3 Solution

The *Packaging* menu under the *Solution* processor contains two commands and two submenus:

Sym BCs

If the problem (geometry and the loading conditions) permits the use of symmetry, the appropriate symmetry boundary conditions can be applied using this function. The user has three options: octant, quarter, and half symmetry.

Fix a Node

The thermal life prediction analysis of packages usually involves simulation of the loading under thermal cycling conditions. The type of loading that the package undergoes is limited to the temperature cycles. After applying the symmetry condition, at least one point (node) in the model must be constrained in all directions in order to avoid rigid-body translations.

Thermal

The load step files for simulation of the thermal cycle for global or submodel analyses are created under this menu. Also, the transition from the nonlinear global model to the nonlinear submodel is achieved herein.

Mechanical

The displacement boundary conditions for the supports and the applied load are specified under this item.

This structure is shown in Fig. 4.56, with submenus of each of the menu items given above. Detailed explanations of each of the menu items (commands) are given in the following sections, with the order given in Figs. 4.57-4.59.

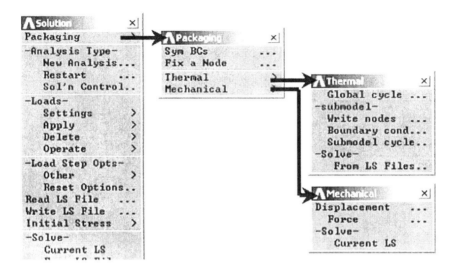

Fig. 4.56 Menu structure of ReliANS under the *Solution* processor in the ANSYS®-GUI.

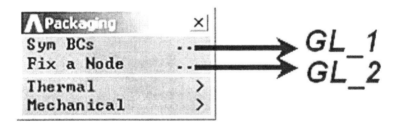

Fig. 4.57 Menu items and their equivalent commands for the *Packaging* menu.

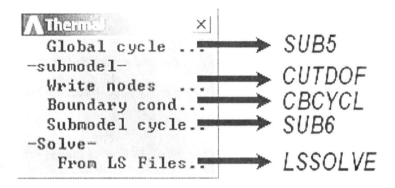

Fig. 4.58   Menu items and their equivalent commands for the *Thermal* submenu.

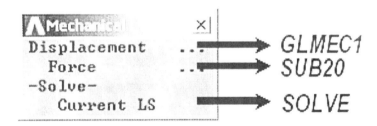

Fig. 4.59   Menu items and their equivalent commands for the *Mechanical* submenu.

4.3.1   Packaging Menu Commands

GL_1, *sym*

This command applies the appropriate symmetry boundary conditions. Figure 4.60 shows the dialog box for this command.

> **sym**   0: octant model
> 1: quarter model
> 2: half model

The menu path for this command is as follows:

```
Solution > Packaging > Sym BCs
```

Notes: The discussion covered in the SJFULL command on symmetry is important and should be reviewed prior to use of the GL_1 command.

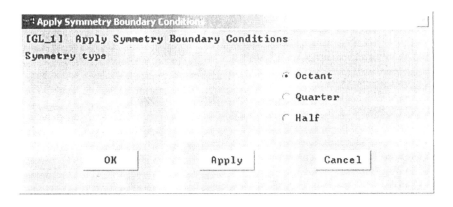

Fig. 4.60   Dialog box for GL_1.

***GL_2**, x, y, z*

This command applies displacement constraints at a specified location in order to prevent rigid-body translations. Figure 4.61 shows the dialog box for this command.

x	*x*-coordinate of the point to be fixed
y	*y*-coordinate of the point to be fixed
z	*z*-coordinate of the point to be fixed

The menu path for this command is as follows:

```
Solution > Packaging > Fix a Node
```

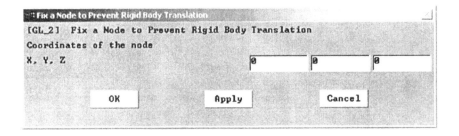

Fig. 4.61   Dialog box for GL_2.

4.3.2   Thermal Submenu Commands

SUB5, *nc, tmx, tmn, t0, trf, dmx, dmn, drup, drdwn, nrup, nrdwn, ndwl*

This command creates the load step files that simulate the thermal cycle for the <u>global model</u>. Figure 4.62 shows the dialog box for this command.

nc	number of cycles (usually 4)
tmx	maximum temperature at dwell
tmn	minimum temperature at dwell
t0	starting temperature (stress-free temperature)
trf	offset temperature (optional)
dmx	duration of dwell at maximum temperature (in seconds)
dmn	duration of dwell at minimum temperature (in seconds)
drup	duration of ramp-up (in seconds)
drdwn	duration of ramp-down (in seconds)
nrup	number of load steps for ramp-up (usually 6 or 7)
nrdwn	number of load steps for ramp-down (usually 6 or 7)
ndwl	number of load steps for dwell (usually 1)

The menu path for this command is as follows:

```
Solution > Packaging > Thermal > Global cycle
```

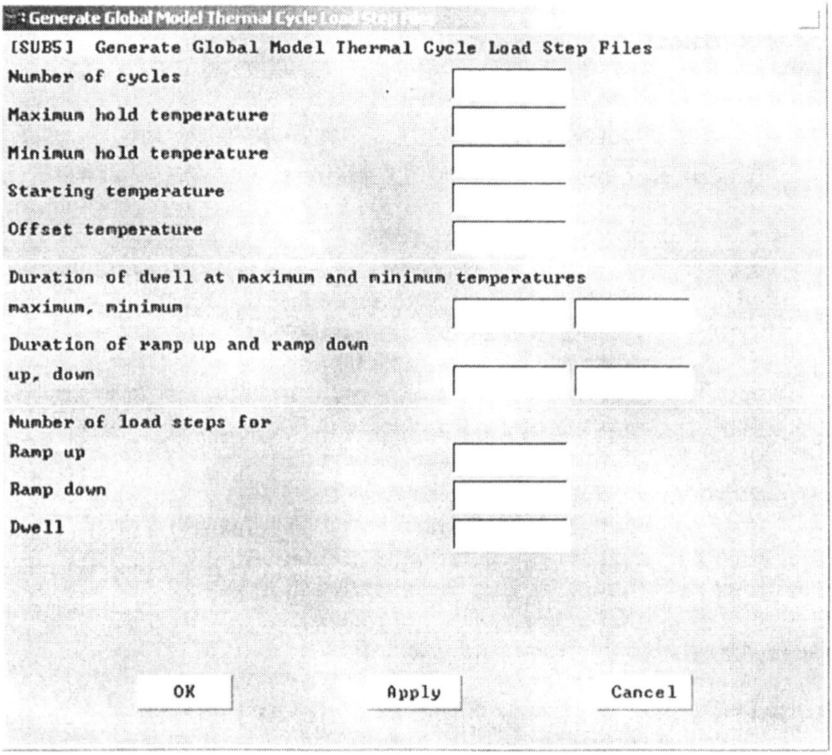

Fig. 4.62 Dialog box for SUB5.

CUTDOF, *usr, y0, y1, x0, x1, z0, z1, sym*

This command creates the *"cutdof.node"* file, which contains the interface nodes between the submodel and the global model. The logic of this operation is exactly the same as that given in the ANSYS® Help System and the user is referred to the "Submodeling" chapter in the ANSYS® Advanced Analysis Guide. Figure 4.63 shows the dialog box for this command.

usr	user-selected nodes option:
	0: Off
	1: On (highly recommended)
y0	y-coordinate of the bottom surface that contains interface nodes between the submodel and the global model
y1	y-coordinate of the top surface that contains interface nodes between the submodel and the global model
x0	x-coordinate of the surface closest to the origin that contains interface nodes between the submodel and the global model
x1	x-coordinate of the surface farthest from the origin that contains interface nodes between the submodel and the global model
z0	z-coordinate of the surface closest to the origin that contains interface nodes between the submodel and the global model
z1	z-coordinate of the surface farthest from the origin that contains interface nodes between the submodel and the global model
sym	symmetry type
	0: octant symmetry
	1: quarter symmetry
	2: half symmetry
	3: no symmetry

The menu path for this command is as follows:

```
Solution > Packaging > Thermal > -submodel- Write nodes
```

Notes: This command writes the interface node information to the *"cutdof.node"* file. Also, it can assist the user in the *selection of nodes* task if the user chooses. Therefore, the user has two options: (1) manual selection of the interface nodes, or (2) utilization of this command. Manual selection is <u>highly recommended</u>. The user must have a working knowledge of the "submodeling" concept within the ANSYS® environment (refer to the help pages mentioned above). Parameters ***y0, y1, x0, x1, z0, z1***, and ***sym*** are disregarded if ***usr*** = 1.

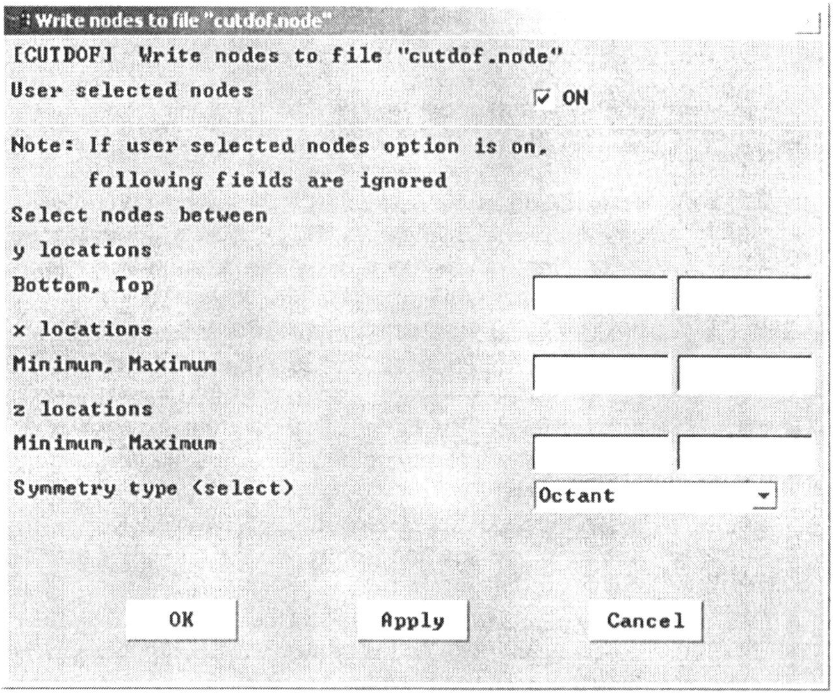

Fig. 4.63   Dialog box for CUTDOF.

CBCYCL, nls, 'gl_name', 'sub_name'

This command reads the "cutdof.node" file (which must be created prior to the execution of this command), interpolates the global model displacement solution for each load step within the thermal cycle, and writes them to the file "cutdof.cbdo." Figure 4.64 shows the dialog box for this command.

nls	number of load steps
'gl_name'	name of global model
'sub_name'	name of submodel

The menu path for this command is as follows:

```
Solution > Packaging > Thermal > -submodel- Boundary
cond
```

Notes: This macro issues the "*/CLEAR*" command twice, which in turn causes ANSYS® to ask the user to verify. The user should click on *Yes* in both cases. This operation is part of the submodeling procedure included in ANSYS® and the user must refer to the "Submodeling" chapter in the ANSYS® Advanced Analysis Guide.

This command requires file names (*jobnames*) for the global and submodels. The user must use *jobnames* while performing this analysis; otherwise, this command will fail. During the execution of this command, ANSYS® will resume "db" files and result files. It is of extreme importance that the user be consistent with the naming convention in ANSYS®. The jobnames for global and submodels must be less than 9 letters.

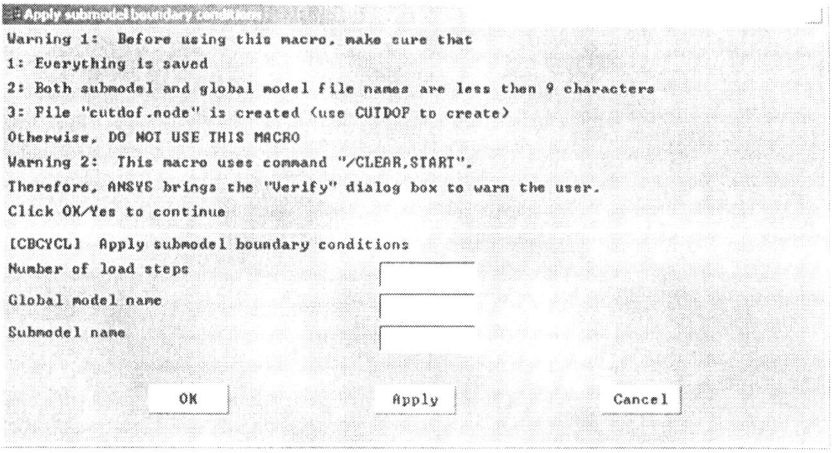

Fig. 4.64   Dialog box for CBCYCL.

SUB6, *nc, tmx, tmn, t0, trf, dmx, dmn, drup, drdwn, nrup, nrdwn, ndwl*

This command creates the load step files that simulate the thermal cycle for the submodel. Figure 4.65 shows the dialog box for this command.

nc	number of cycles (same as that for SUB5)
tmx	maximum temperature at dwell
tmn	minimum temperature at dwell
t0	starting temperature (stress-free temperature)
trf	offset temperature (optional)
dmx	duration of dwell at maximum temperature (in seconds)
dmn	duration of dwell at minimum temperature (in seconds)
drup	duration of ramp-up (in seconds)
drdwn	duration of ramp-down (in seconds)
nrup	number of load steps for ramp-up (same as that for SUB5)
nrdwn	number of load steps for ramp-down (same as that for SUB5)
ndwl	number of load steps for dwell (same as that for SUB5)

The menu path for this command is as follows:

```
Solution> Packaging > Thermal > -submodel- Submodel
cycle
```

Notes: This command searches for the load step files that are created by CBCYCL with the same name as that of the submodel *jobname*. All the parameters specified in this command must be the same as the ones specified in the global model.

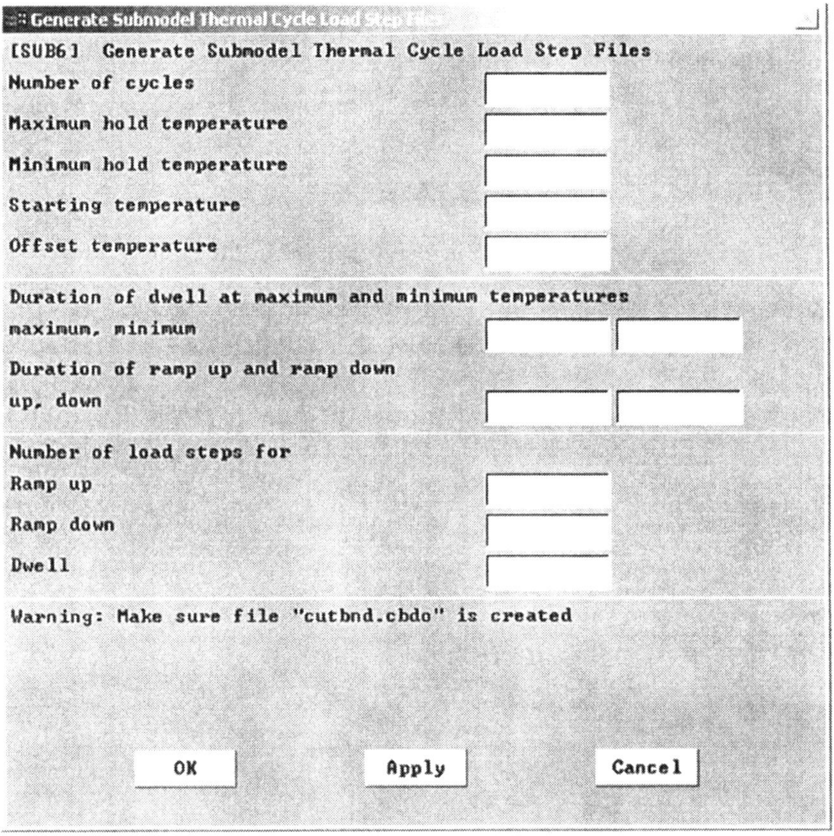

Fig. 4.65 Dialog box for SUB6.

***LSSOLVE**, Ismin, Ismax, Isinc*

This command starts the solution procedure for the global or submodel. It is an ANSYS® command, and it is included under the *Packaging* menu strictly for convenience. Figure 4.66 shows the dialog box for this command.

Ismin   starting load step file number
Ismax   ending load step file number
Isinc   file number increment

The menu path for this command is as follows:

Solution > Packaging > Thermal > -Solve- From LS Files

Notes: Prior to issuing this command, the user must know the exact number of load step files. This can be found by exploring the working directory. The load step files have the following format: *jobname*.s##. Here, ## is the load step number, i.e., for load step 1 the file name is *jobname*.s01, for load step 24 it is *jobname*.s24, etc. The solution normally starts from load step 1 (*Ismin* = 1) until the last one, with an increment of 1 (*Isinc* = 1).

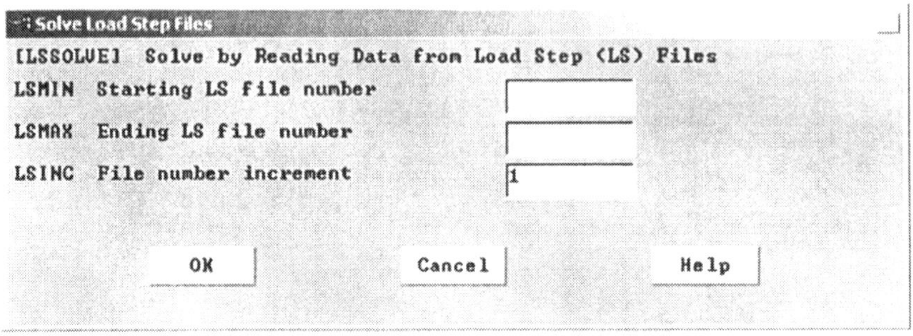

Fig. 4.66   Dialog box for LSSOLVE.

4.3.3   Mechanical Submenu Commands

GLMEC1, x, y, z, rad, sym

This command applies the displacement constraints to nodes corresponding to the support locations. Figure 4.67 shows the dialog box for this command.

x	x-coordinate of the center point of the support
y	y-coordinate of the center point of the support
z	z-coordinate of the center point of the support
rad	radius of the support
sym	symmetry type
	0: half symmetry
	1: no symmetry

The menu path for this command is as follows:

Solution > Packaging > Mechanical > Displacement

Notes: This command assumes that the support is at the corners of an equidistant triangle. Therefore, the coordinates given in this command do <u>not</u> correspond to the coordinates of the nodes that are subjected to the displacement constraints. They correspond to the center of the equidistant triangle. The actual support locations and the nodes closest to these locations are evaluated by this command.

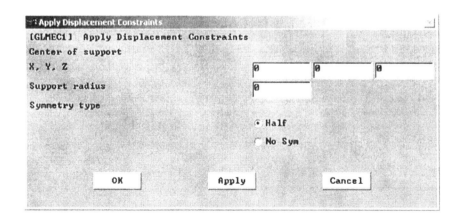

Fig. 4.67   Dialog box for GLMEC1.

SUB20, *force, x, y, z, rad, sym*

This command facilitates the application of the external load. Figure 4.68
shows the dialog box for this command.

force	applied normal force parallel to the y-coordinate
x	x-coordinate of the center of the area of loading
y	y-coordinate of the center of the area of loading
z	z-coordinate of the center of the area of loading
rad	radius of the area of loading
sym	symmetry type
	0: half symmetry
	1: no symmetry

The menu path for this command is as follows:

Solution > Packaging > Mechanical > Force

Notes: The externally applied force is simulated by using this command.
This command obtains the number of nodes at the area of loading, divides
the total force into that many nodes, and applies the resulting value as point
loads.

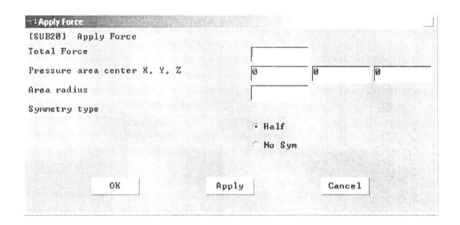

Fig. 4.68   Dialog box for SUB20.

4.4 Postprocessing

The *Packaging* menu under the *General Postprocessor* contains four menu items:

Thermal Life   This menu item calculates the volume-weighted-average plastic work density for the selected elements and the corresponding predicted life.

Eqv strs & strn   This menu item calculates the volume-weighted-average von Mises stresses and strains.

Assembly Stfns   This menu item calculates the effective stiffness, imposed strain, and assembly stiffness based on the values of volume-weighted-average von Mises stresses and strains.

Bending Life   This menu item calculates the strain energy per cycle using 1-D combined creep and time-independent plasticity and the corresponding predicted life.

This structure is shown in Figure 4.69, with the menu items given above and their equivalent commands.

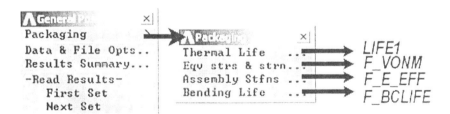

Fig. 4.69   Menu structure of ReliANS under the *General Processor* in the ANSYS®-GUI.

4.4.1   Thermal Life Menu Item

LIFE1, *t, nls, cyc, r, usr, ybot, ytop, mat*

This command calculates the predicted fatigue life based on the volume-weighted-average plastic work for the selected element set. Figure 4.70 shows the dialog box for this command.

t	element thickness
	0: 0.5 mil
	1: 1.0 mil
	2: 1.5 mils
nls	total number of load steps
cyc	total number of cycles
r	radius of cross-section where crack initiation and propagation is expected
usr	user-selected elements option
	0: Off
	1: On (highly recommended)
ybot	bottom *y*-coordinate for elements to be selected (disregarded if *usr* = 1)
ytop	top *y*-coordinate for elements to be selected (disregarded if *usr* = 1)
mat	material number for solder (default is 1)

The menu path for this command is as follows:

```
General Postproc > Packaging > Thermal Life
```

Notes: The life-prediction method developed by Darveaux (2000) provides empirical constants that are dependent on the element thickness (in the *y*-direction) used at the critical solder joint portions. These constants are available for element sizes 0.5, 1.0, and 1.5 mils. ReliANS commands under the preprocessor provide the user with the opportunity to create a mesh that exactly matches these element thicknesses. Therefore, the user must know (preferably create the mesh according to these thicknesses) the thickness information prior to the use of this command.

The volume-weighted-average plastic work is calculated in a time interval, usually the last cycle. However, ReliANS gives the user the opportunity to review the value of this quantity at any one of the previous cycles. This is achieved by specifying the total number of load steps (*nls*) and the total number of cycles (*cyc*) corresponding to the cycle of interest. For example, if 4 cycles are simulated using 60 load steps, the user must specify *nls* = 60 and *cyc* = 4 in order to obtain the predicted fatigue life based on the volume-weighted-average plastic work during the 4th cycle. On the other hand, if the user is interested in the quantities associated with the 3rd cycle, the corresponding information should be given as *nls* = 45 and *cyc* = 3.

It is recommended that the user select the elements for which the volume-weighted-average plastic work is to be calculated.

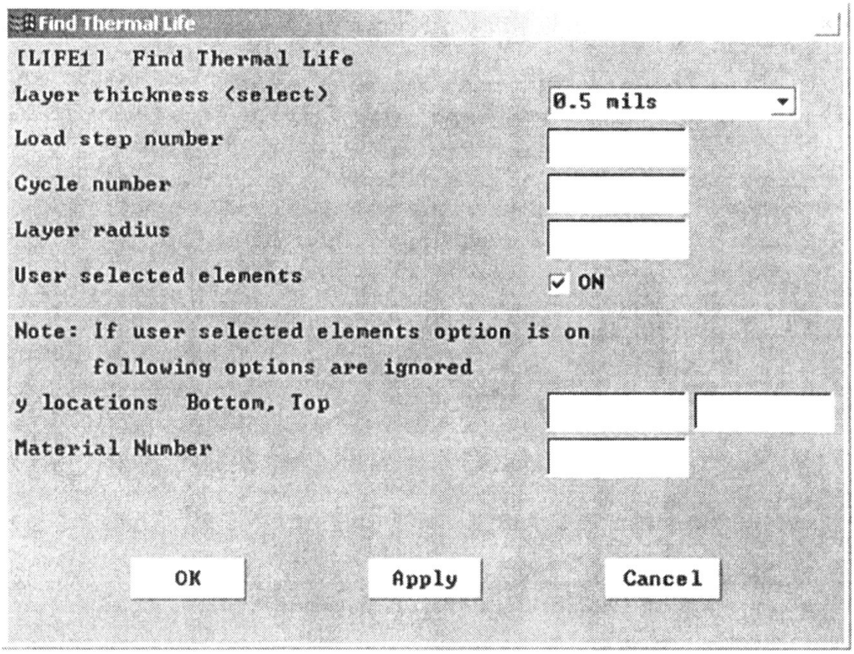

Fig. 4.70 Dialog box for LIFE1.

4.4.2   Equivalent Stress and Strain Menu Item

F_VONM, *usr, pitch, x, z, ybot, ytop, mat*

This command calculates the volume-weighted-average von Mises stresses and strains for the selected element set. Figure 4.71 shows the dialog box for this command.

usr	user-selected elements option
	0: Off
	1: On (highly recommended)
pitch	pitch
x	x-coordinate of the desired solder joint
z	z-coordinate of the desired solder joint
ybot	bottom y-coordinate for elements to be selected (disregarded if **usr** = 1)
ytop	top y-coordinate for elements to be selected (disregarded if **usr** = 1)
mat	material number for solder (default is 1)

The menu path for this command is as follows:

```
General Postproc > Packaging > Eqv strs & strn
```

Note: It is strongly recommended that the user select the elements for this calculation.

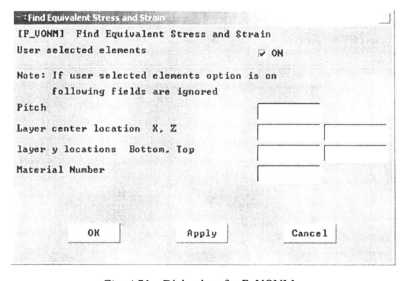

Fig. 4.71   Dialog box for F_VONM.

4.4.3 Assembly Stiffness Menu Item

F_E_EFF, s1, e1, s01, e01, e_sol

This command calculates the effective stiffness, imposed strain, and assembly stiffness based on the volume-weighted-average von Mises stresses and strains calculated by F_VONM. Figure 4.72 shows the dialog box for this command.

s1	volume-weighted-average von Mises stress for the selected elements when the actual elastic modulus for solder is used (MPa)
e1	volume-weighted-average von Mises strain for the selected elements when the actual elastic modulus for solder is used
s001	volume-weighted-average von Mises stress for the selected elements when the elastic modulus for solder is reduced by 100 (MPa)
e001	volume-weighted-average von Mises strain for the selected elements when the elastic modulus for solder is reduced by 100
e_sol	elastic modulus for solder (psi)

The menu path for this command is as follows:

```
General Postproc > Packaging > Assembly Stfns
```

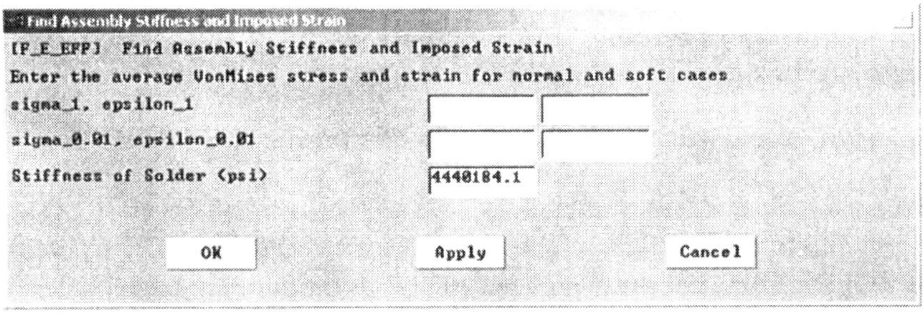

Fig. 4.72   Dialog box for F_E_EFF.

4.4.4 Mechanical Life Menu Item

F_BCLIFE, *c5, a1, n, qa, k, t, et, bt, g0, g1, m, c6, s, e0, p, a, nc, ni*

This command calculates the predicted fatigue life based on the strain energy calculated using a 1-D combined creep and time-independent plasticity simulation. Figure 4.73 shows the dialog box for this command.

c5	material constant C_{5t} in Eq. (1.11)
a1	material constant α_{1t} in Eq. (1.11)
n	material constant n in Eq. (1.11)
qa	material constant Q_a in Eq. (1.11)
k	material constant k in Eq. (1.11)
t	absolute temperature T in Eq. (1.11)
et	material constant ε_T in Eq. (1.10)
bt	material constant B_t in Eq. (1.10)
g0	material constant G_0 used for the calculation of shear modulus in Eq. (1.13)
g1	material constant G_1 used for the calculation of shear modulus in Eq. (1.13)
m	material constant m in Eq. (1.13)
c6	material constant C_{6t} in Eq. (1.13)
s	assembly stiffness (S) calculated by F_E_EFF (psi)
e0	imposed strain (ε_0) calculated by F_E_EFF
p	period of the cyclic loading
a	crack path length (diameter of the selected elements)
nc	number of cycles to be simulated
ni	number of time intervals to be used for each cycle

The menu path for this command is as follows:

```
General Postproc > Packaging > Mechanical Life
```

Notes: Default values of all the material constants are included in this command. Prior to this command, if the user executes the F_E_EFF command to find the imposed strain and assembly stiffness, their computed values appear in this menu.

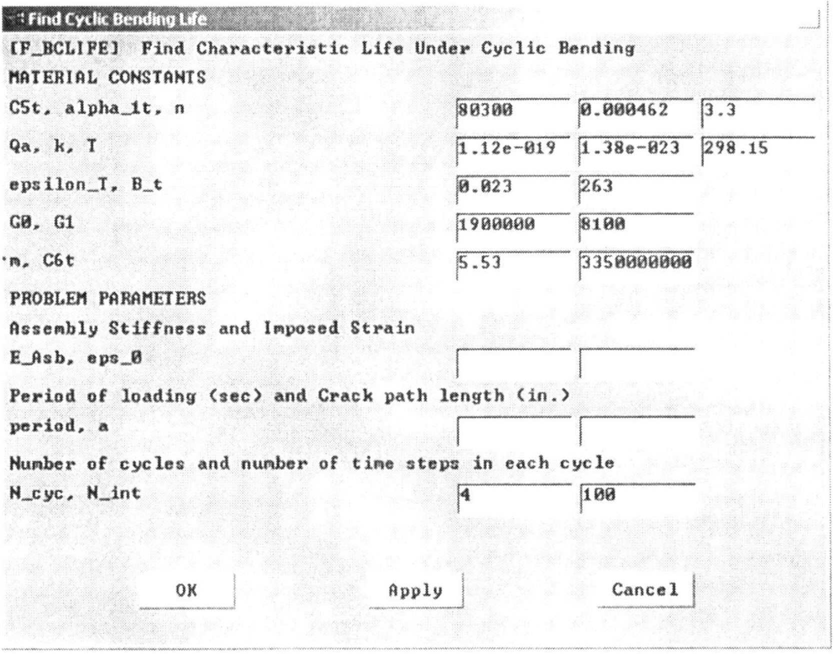

Fig. 4.73 Dialog box for F_BCLIFE.

4.5 Reference

Darveaux, R., 2000, "Effect of Simulation Methodology on Solder Joint Crack Growth Correlation," *Proceedings, 50th Electronic Components and Technology Conference*, IEEE, New York, pp. 1048-1058.

Appendix A

INSTALLATION AND EXECUTION

The installation and maintenance of ReliANS, add-on software to ANSYS®, is **dependent** on the computer environment. The SETUP program will install ReliANS automatically for Windows 98, ME, 2000, XP, and NT environments on a PC platform. The SETUP program will *not* install ReliANS on UNIX platforms. However, it is rather straightforward to perform this task "manually." Except for the one-dimensional combined creep and time-independent plasticity simulation capability, all other macros are available for UNIX platforms. For this task, an executable file is called from within ANSYS using the "/sys" command on PC platforms.

The user **must** know the path for the ANSYS® installation directory prior to starting the installation procedure. This directory can be on the network or on a local hard drive. The installation program for ReliANS needs read-only access to the ANSYS® installation directory (no modifications will be made to the original ANSYS® installation). Without this information, the installation will **fail**. Users are advised to contact their system administrator to obtain this information. During the installation, in addition to the ANSYS® installation directory, the user will be asked to provide paths for the ReliANS *Macro directory* and *working directories* from which these macros will be accessible. These directories should be created on the user's local hard drive. Although it is not necessary, it is advised that the user create working directories prior to the installation. Otherwise, the user will have to create these directories during the installation process. Note that the user will have access to the ReliANS add-on package only from the *working directories* specified during the installation. However, the process to add new directories from which the add-on package may be used is very straightforward.

The add-on package comes on a CD-ROM. The steps for the installation and adding of new directories are given below.

A.1 Steps for ReliANS Installation on a PC Platform

- Place the installation CD-ROM into the appropriate drive.
- Start (double-click) *SETUP.exe* in the root directory of the CD-ROM.
- Choose Install.
- Browse and select an ANSYS® installation directory.
- Browse and select the ReliANS installation directory.
- Browse and select the working directories from which ReliANS will be accessible.
- Finish.

A.2 Steps for Adding Working Directories for Use in ReliANS

- Place the installation CD-ROM into the appropriate drive.
- Start (double-click) *SETUP.exe* in the root directory of the CD-ROM.
- Choose *Add Working Directory*.
- Browse and select the ReliANS installation directory.
- Browse and select the working directory from which ReliANS will be accessible.
- Finish.

The steps discussed above for installation and adding working directories are demonstrated in detail next.

A.3 Demonstration of ReliANS Installation and Adding New Directories

STEP 1:   Place the installation CD-ROM into the appropriate drive.

STEP 2:   Start (double-click) *SETUP.exe* in the root directory of the CD-ROM. Upon starting, the initial window for the installation procedure will appear, as shown in Fig. A.1. Click on **Next**. The new window, shown in Fig. A.2, gives the names of the developers of ReliANS. Click on **Next**.

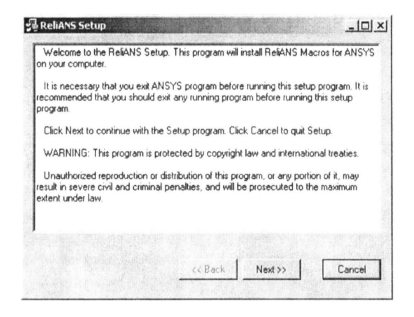

Fig. A.1   Initial SETUP window.

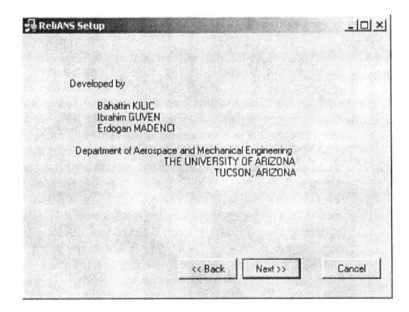

Fig. A.2   SETUP window following the initial window.

STEP 3:   Indicate whether to install or add new directory. To install, click
on the radio button next to **Install** and hit **Next** (refer to Fig.
A.3). To add new working directories, click on the radio button
next to **Add working directory** and hit **Next** (refer to Fig. A.3),
and continue with STEP 5.

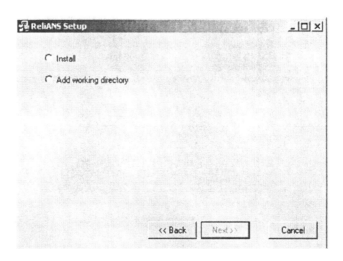

Fig. A.3   SETUP window for selecting installation or the
addition of a working directory.

STEP 4:   Browse and select ANSYS® installation directory. The user
must provide the path for the ANSYS® installation directory.
Browse, select, and hit **Save**. ANSYS® versions earlier than 5.7
will typically be installed in a directory located in the root
directory of the drive. For example, the ANSYS® 5.6 installation directory has a path similar to "C:\ANSYS56," whereas
ANSYS® 5.7 is installed under "C:\Program Files\Ansys Inc."
The SETUP program will perform a search of the user's hard
drive and find the ANSYS® installation directory, as shown in
Fig. A.4 (valid for ANSYS® versions 5.7 and higher). If only
one ANSYS® version is installed on the user's computer, then
selecting the "C:\Program Files\Ansys Inc" folder is sufficient
for the installation of ReliANS. If there are more than one
ANSYS® versions installed, then the user must select the
particular ANSYS® version to be used with ReliANS (shown
for ANSYS® 5.7 in Fig. A.5). Once the correct folder is
selected, click on **Save**. The selected ANSYS® installation
directory will be shown to the user in the window, as shown in
Fig. A.6. Click on **Next**.

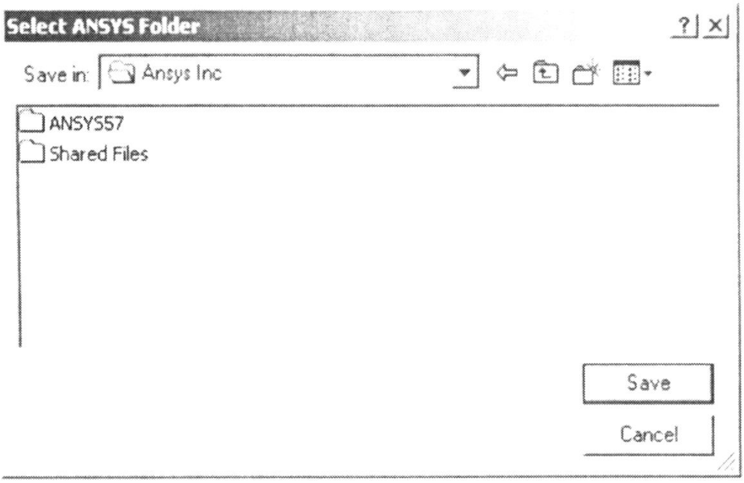

Fig. A.4 SETUP window for selecting ANSYS® installation folder.

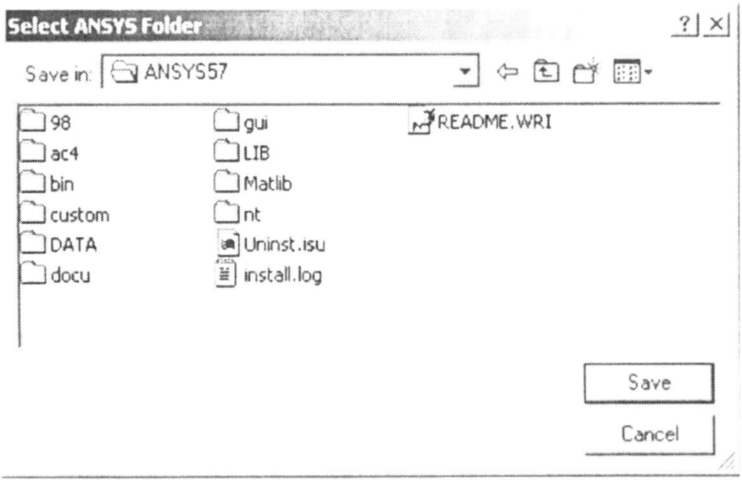

Fig. A.5 SETUP window for selecting desired ANSYS® version installation folder if there is more than one version installed.

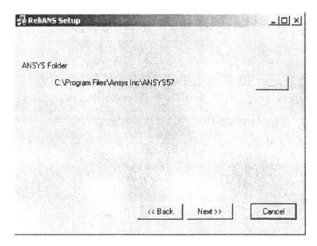

Fig. A.6   SETUP window after selecting ANSYS® installation folder.

STEP 5:   <u>Browse and select the ReliANS installation directory.</u> After the
selection of the ANSYS® installation directory, SETUP will ask
the user to indicate the directory for ReliANS Macro storage, as
shown in Fig. A.7. At this point, SETUP will suggest an installlation directory. The user can browse for an alternate location;
however, because of ANSYS® program limitations, the path for
this location **must not** have any spaces and **must not** exceed 64
characters. Once finished with this step, click on **Next**.

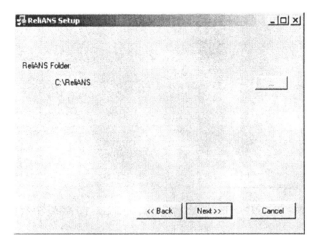

Fig. A.7. SETUP window for selecting ReliANS installation folder.

STEP 6:   Browse and select the *working directory* from which ReliANS
                will be accessible. After selection of the ReliANS installation
                directory, the window shown in Fig. A.8 will appear. By click-
                ing on the **Add Directory** button, a browse window will appear
                (shown in Fig. A.9). Browse and select the previously created
                working directory. If the *working directory* was not created pre-
                viously, it can be created by clicking on the *Create New Folder*
                icon in the browse window. Once the *working directory* has
                been selected, click on **Save**. The SETUP window will appear,
                as shown in Fig. A.10, with the path for the working directory.
                The user can add more working directories by clicking on the
                Add Directory button or finalize the installation by clicking on
                the **Next** button. Once the Next button is selected, the final
                window for the ReliANS setup will appear, as shown in Fig.
                A.11, listing the working directories added in the current
                installation session. Click on **Finish** to exit SETUP.

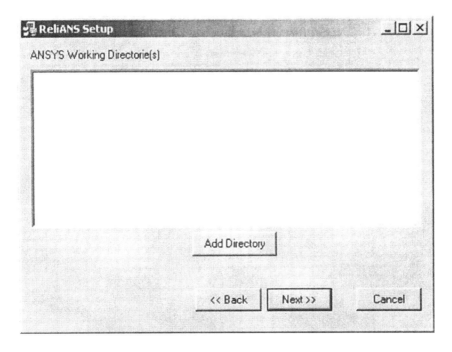

Fig. A.8   SETUP window for adding working directories.

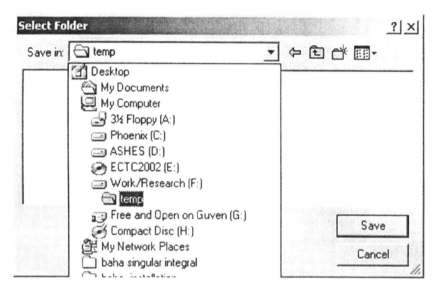

Fig. A.9   SETUP window while browsing during the addition of
            working directories.

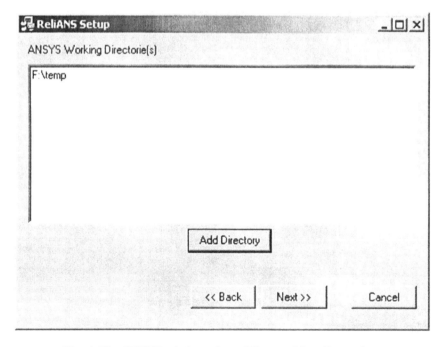

Fig. A.10   SETUP window after adding working directories.

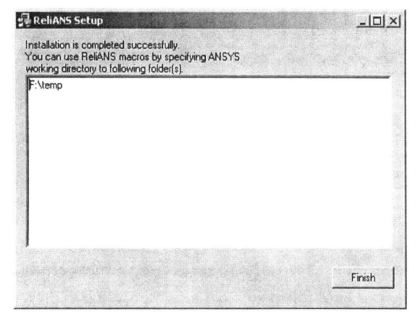

Fig. A.11 Final SETUP window.

A.4 Installation of ReliANS on UNIX Systems

As mentioned previously, the installation of ReliANS on UNIX platforms is performed "manually," and also the one-dimensional combined creep and time-independent plasticity simulation capability of ReliANS is *not* supported on a UNIX platform because of the incompatibilities arising from the different versions of UNIX operating systems for different servers.

For "manual" installation, the user must have access to a PC. Step-by-step instructions are given below.

1. Locate the files "menulist**.ans" and "UIMENU.GRN" in the UNIX system. The characters "**" in "menulist**.ans" correspond to the version of ANSYS® (e.g., for ANSYS® 5.7, this file is called "menulist57.ans").

   Example location for "menulist57.ans" file:

   /usr/local/ansys57/ansys57/docu/english

   Example location for "UIMENU.GRN" file:

   /usr/local/ansys57/ansys57/docu/English/UIDL

2. Create a temporary folder on the PC that will serve as "dummy" installation directory for ANSYS®.

   Example location for "dummy" installation directory for ANSYS®:

   C:\ans_temp

3. Transfer the "menulist**.ans" and "UIMENU.GRN" files from the UNIX system to the folder created in Step 2. This is done so that the SETUP program can create the graphic user interface (GUI) files properly.

4. Create another "dummy" folder in the PC as the ANSYS® working directory from which ReliANS will be accessible (e.g.. "C:\wd_ReliANS").

5. Place the installation CD-ROM into the appropriate drive of the PC.

6. Start (double-click) *SETUP.exe* in the root directory of the CD-ROM.

7. Choose Install.

8. Browse and select the "dummy" ANSYS® installation directory created in Step 2 (e.g., "C:\ans_temp").

9. Browse and select the ReliANS installation directory (the SETUP program will create this directory as "C:\ReliANS" by default).

10. Browse and select the "dummy" working directory created in Step 4.

11. Finish.

12. Notice that several files are copied in the ReliANS installation directory ("C:\ReliANS" by default). Also, notice that the "dummy" working directory created in Step 4 contains the "menulist**.ans" file.

13. Create the ReliANS installation directory in the UNIX system.

    Example location for ReliANS installation directory in the UNIX system:

    /home/u7/user1/ReliANS

14. Transfer all the files in the ReliANS installation folder on the PC to the ReliANS directory on the UNIX system (created in previous step) in ASCII mode. The user must pay special attention to the file transfer mode as some file transfer programs modify the ASCII files when they are transferred from PC to UNIX, causing a read failure in ANSYS®.

15. Create the ANSYS® working directory from which ReliANS will be accessible on the UNIX system.

    Example location for ANSYS® working directory from which ReliANS will be accessible on the UNIX system:

    /home/u7/user1/eptool

16. Transfer the "menulist**.ans" file from the "dummy" working directory on PC to the working directory on UNIX system. The file must be transferred in ASCII mode as explained in Step 14.

17. Open the "menulist**.ans" file on the UNIX system with a text editor. Notice that the last two rows are similar to:

    C:\ReliANS\EPMENU.GRN
    C:\ReliANS\EPFUNC.GRN

    Modify these lines as (following the example naming convention given in Step 13):

    /home/u7/user1/ReliANS/EPMENU.GRN
    /home/u7/user1/ReliANS/EPFUNC.GRN

    Save and exit.

18. Open the "EPMENU.GRN" file in the ReliANS installation directory on the UNIX system with a text editor. The following command is issued at three locations in this file:

    :C)/psearch,'C:\\ReliANS'

    In all three locations, modify the path given between the quotation marks as (following the example naming convention given in Step 13):

    :C)/psearch,'/home/u7/user1/ReliANS'

    Save and exit.

19. The ReliANS software is now ready for use. If additional working directories are desired, simply copy the "menulist**.ans" file to the new working directory without any modifications.

Appendix B

INPUT LISTINGS FOR CASE STUDIES

B.1 Input Listing for the Case Study Given in Chapter 2

```
FINISH                                                  ! begin level
/CLEAR,START                                            ! clear database
/FILENAME,DEMO
/TITLE,Thermal Fatigue Reliability Analysis - Global
/PSEARCH,C:\ReliANS                         ! set the macro directory
/PREP7                                         ! enter preprocessor
ET,1,45                                 ! element type # 1 must be solid45
ET,2,42                                 ! element type # 2 must be plane42

! enter solder joint information
SJGEO1,2,0.146,0.235,0.15,0.146,0.232,0.021,0.06,0.067,,,,,
SJGEO2,7,0.174,0.029,0.188,0.058,0.192,0.087,0.19,0.116,0.185,0.145,
       (continued)                          0.177,0.174,0.165,0.203
SJGEO3,3,2,,1,2,2,,3,,8,1,1,,

! create solder joint array
SJFULL,2.75,0.692,0.25,0.5,2,7,1,2,0

! enter motherboard information
FR4GEO1,9,0.05,0.018,0.2,0.018,0.12,0.018,0.2,0.018,0.05
FR4GEO2,9,1,1,1,1,1,1,1,1,1
FR4GEO3,9,5,2,4,3,4,2,4,3,5

! create mother board
BLCK7,2.75,0.0,0.25,0.5,2,7,0.15,,2,,,8,0
BLCK9,3.5,0.0,0.0,6.5,10,45,5,14
BLCK10,0,0,0,2.5,10

! create layers above solder joint level
BLCK2,2.75,0.972,0.25,0.5,2,7,0.146,0.235,0.012,60,2,1,1,8,6,0
BLCK4,3.5,0.972,0.0,0.5,0.012,4,45,2,1,14,6
BLCK5,0.0,0.972,0.0,2.5,0.012,10,1,6

PARAM1=0.146+0.012/TAN(ATAN(1)*4/3)
BLCK2,2.75,0.897,0.25,0.5,2,7,  PARAM1,0.235,0.075,60,2,1,2,8,7,0
BLCK4,3.5,0.897,0.0,0.5,0.075,4,45,2,2,14,7
BLCK5,0.0,0.897,0.0,2.5,0.075,10,2,7

BLCK2,2.75,0.984,0.25,0.5,2,7,0.235,,0.021,90,3,1,1,8,8,0
BLCK4,3.5,0.984,0.0,0.5,0.021,4,45,2,1,14,8
BLCK5,0.0,0.984,0.0,2.5,0.021,10,1,8
```

```
BLCK1,2.75,1.005,0.25,0.5,2,7,0.146,0.235,0.027,3,2,1,1,8,8,0
BLCK4,3.5,1.005,0.0,0.5,0.027,4,45,2,1,14,8
BLCK5,0.0,1.005,0.0,2.5,0.027,10,1,8
BLCK5,0.0,1.032,0.0,2.5,0.07,10,1,9
BLCK5,0.0,1.102,0.0,2.5,0.3,10,2,11
BLCK1,2.75,1.032,0.25,0.5,2,7,0.146,0.235,0.07,3,2,1,1,8,12,0
BLCK4,3.5,1.032,0.0,0.5,0.07,4,45,2,1,14,12
BLCK1,2.75,1.102,0.25,0.5,2,7,0.146,0.235,0.3,3,2,1,2,8,12,0
BLCK4,3.5,1.102,0.0,0.5,0.3,4,45,2,2,14,12
BLCK1,2.75,1.402,0.25,0.5,2,7,0.146,0.235,0.36,3,2,1,2,8,12,0
BLCK4,3.5,1.402,0.0,0.5,0.36,4,45,2,2,14,12
BLCK5,0.0,1.402,0.0,2.5,0.36,10,2,12

! change the solder elements from SOLID45 to VISCO107
CHGVISC,1

! set material properties
EPMAT,1,26                                            ! solder
EPMAT,2,14                                            ! copper
EPMAT,3,14                                ! copper in motherboard
EPMAT,4,11                                               ! FR4
EPMAT,5,30                                         ! polyimide
EPMAT,6,31                                          ! adhesive
EPMAT,7,30                                         ! substrate
EPMAT,8,29                                        ! solder mask
EPMAT,9,28                                                ! die
EPMAT,11,15                                       ! die-attach
EPMAT,12,27                                              ! mold

SAVE

/SOLU                                     ! enter solution processor

! specify displacement constraints
GL_1,0 ! set symmetry type (octant)
GL_2,0,0,0                       ! fix a node to prevent rigid body motion

! create load step files simulating the thermal cycle
SUB5,4,373.15,233.15,298.15,0.0,1560,1440,240,360,6,6,1

ALLSEL,ALL                                       ! select everything

SAVE

LSSOLVE,1,60                             ! solve from load step files

SAVE

FINISH

! the user examines the accumulated plastic work and decides on the
! most critical solder joint - the x and z coordinates of the center
! of the most critical solder joint is needed for the submodel

/CLEAR,START
/FILENAME,DEMOSUB
/TITLE,Thermal Fatigue Reliability Analysis - Submodel
/PSEARCH,C:\ReliANS                          ! set the macro directory

/PREP7                                        ! enter preprocessor
ET,1,45                          ! element type # 1 must be solid45
ET,2,42                          ! element type # 2 must be plane42
```

```
! create the most critical solder joint
SJGEO1,2,0.146,0.235,0.15,0.146,0.232,0.021,0.06,0.067,,1,0.0127,,
SJGEO2,7,0.174,0.029,0.188,0.058,0.192,0.087,0.19,0.116,0.185,0.145,
        (continued)                     0.177,0.174,0.165,0.203
SJGEO3,6,3,,2,2,2,,8,,16,,,,
SJFULL,3.25,0.692,3.25,0.5,1,1,1,2,0

! create motherboard
FR4GEO1,3,0.1,0.018,0.05
FR4GEO2,3,3,1,2
FR4GEO3,3,4,3,5
BLCK7,3.25,0.524,3.25,0.5,1,1,0.15,,4,,4,16,0

! create layers above solder joint level
BLCK2,3.25,0.972,3.25,0.5,1,1,0.146,0.235,0.012,60,3,3,1,16,6,0

PARAM1=0.146+0.012/TAN(ATAN(1)*4/3)
BLCK2,3.25,0.897,3.25,0.5,1,1,PARAM1,0.235,0.075,60,3,3,3,16,7,0
BLCK2,3.25,0.984,3.25,0.5,1,1,0.235,,0.021,90,6,3,2,16,8,0
BLCK1,3.25,1.005,3.25,0.5,1,1,0.146,0.235,0.027,6,3,3,2,16,8,0
BLCK1,3.25,1.032,3.25,0.5,1,1,0.146,0.235,0.1,6,3,3,4,16,12,0

! change the solder elements from SOLID45 to VISCO107
CHGVISC,1

! set material properties
EPMAT,1,26                                              ! solder
EPMAT,2,14                                              ! copper
EPMAT,3,14                                   ! copper in motherboard
EPMAT,4,11                                                 ! FR4
EPMAT,5,30                                             ! polyimide
EPMAT,6,31                                              ! adhesive
EPMAT,7,30                                             ! substrate
EPMAT,8,29                                            ! solder mask
EPMAT,12,27                                                ! mold

! write interface nodes between the global and the submodel into file
CUTDOF,0,0.524,1.132,3.0,3.5,3.0,3.25,0

ALLSEL,ALL
SAVE

! read in and interpolate global model results for the submodel nodes
! on the interface between the global and submodel
CBCYCL,60,'DEMO','DEMOSUB'

/SOLU

! create load step files for the submodel simulating the thermal
! cycles
SUB6,4,373.15,233.15,298.15,0.0,1560,1440,240,360,6,6,1

ALLSEL,ALL
SAVE
LSSOLVE,1,60
SAVE
FINISH
/POST1
LIFE1,0,60,4,0.146,0,0.9713,0.984,1               ! calculate fatigue life
```

B.2 Input Listing for the Case Study Given in Chapter 3

```
FINISH
/CLEAR,START
/FILENAME,EXAMPLE2
/TITLE,Mechanical Bending Fatigue Reliability Analysis
/PSEARCH,C:\ReliANS                              ! set the macro directory
/PREP7
ET,1,45
ET,2,42

! create solder joints
SJGEO1,2,0.182,0.27,0.15,0.13,0.3231,0.0317,,0.069,,2,0.0127,2,0.0127
SJGEO2,6,0.176,0.0404,0.208,0.0808,0.228,0.1212,0.234,0.1616,0.232,
       (continued)                    0.2019,0.222,0.2423,0.208,0.2827
SJGEO3,3,1,,1,2,,,5,1,8,1,1,,
SJFULL,0.4,0.692,3.6,0.8,7,3,1,2,2
SJFULL,0.4,0.692,-5.2,0.8,7,3,1,2,2
SJFULL,3.6,0.692,-2.8,0.8,3,8,1,2,2

! create solder mask
BLCK2,0.4,1.0151,3.6,0.8,7,3,0.27,,0.0317,90,3,,1,8,8,2
BLCK2,0.4,1.0151,-5.2,0.8,7,3,0.27,,0.0317,90,3,,1,8,8,2
BLCK2,3.6,1.0151,-2.8,0.8,3,8,0.27,,0.0317,90,3,,1,8,8,2

BLCK3,0,1.0151,-3.2,3.2,0.0317,6.4,8,1,16,8
BLCK3,0,1.0151,5.6,5.6,0.0317,0.4,14,1,1,8
BLCK3,0,1.0151,-6,5.6,0.0317,0.4,14,1,1,8
BLCK3,5.6,1.0151,-5.6,0.4,0.0317,11.2,1,1,28,8
BLCK3,5.6,1.0151,-6,0.4,0.0317,0.4,1,1,1,8
BLCK3,5.6,1.0151,5.6,0.4,0.0317,0.4,1,1,1,8

BLCK1,0.4,1.0468,3.6,0.8,7,3,0.182,0.27,0.0163,3,1,2,1,8,8,2
BLCK1,0.4,1.0468,-5.2,0.8,7,3,0.182,0.27,0.0163,3,1,2,1,8,8,2
BLCK1,3.6,1.0468,-2.8,0.8,3,8,0.182,0.27,0.0163,3,1,2,1,8,8,2

BLCK3,0,1.0468,-3.2,3.2,0.0163,6.4,8,1,16,8
BLCK3,0,1.0468,5.6,5.6,0.0163,0.4,14,1,1,8
BLCK3,0,1.0468,-6,5.6,0.0163,0.4,14,1,1,8
BLCK3,5.6,1.0468,-5.6,0.4,0.0163,11.2,1,1,28,8
BLCK3,5.6,1.0468,-6,0.4,0.0163,0.4,1,1,1,8
BLCK3,5.6,1.0468,5.6,0.4,0.0163,0.4,1,1,1,8

! create substrate
BLCK2,0.4,0.9401,3.6,0.8,7,3,0.182,0.27,0.075,50,1,2,1,8,6,2
BLCK2,0.4,0.9401,-5.2,0.8,7,3,0.182,0.27,0.075,50,1,2,1,8,6,2
BLCK2,3.6,0.9401,-2.8,0.8,3,8,0.182,0.27,0.075,50,1,2,1,8,6,2

BLCK3,0,0.9401,-3.2,3.2,0.075,6.4,8,1,16,6
BLCK3,0,0.9401,5.6,5.6,0.075,0.4,14,1,1,6
BLCK3,0,0.9401,-6,5.6,0.075,0.4,14,1,1,6
BLCK3,5.6,0.9401,-5.6,0.4,0.075,11.2,1,1,28,6
BLCK3,5.6,0.9401,-6,0.4,0.075,0.4,1,1,1,6
BLCK3,5.6,0.9401,5.6,0.4,0.075,0.4,1,1,1,6

! create die attach
BLCK1,0.4,1.0631,3.6,0.8,7,3,0.182,0.27,0.07,3,1,2,1,8,10,2
BLCK1,0.4,1.0631,-5.2,0.8,7,3,0.182,0.27,0.07,3,1,2,1,8,10,2
BLCK1,3.6,1.0631,-2.8,0.8,3,8,0.182,0.27,0.07,3,1,2,1,8,10,2
```

```
BLCK3,0,1.0631,-3.2,3.2,0.07,6.4,8,1,16,10
BLCK3,0,1.0631,5.6,5.6,0.07,0.4,14,1,1,10
BLCK3,0,1.0631,-6,5.6,0.07,0.4,14,1,1,10
BLCK3,5.6,1.0631,-5.6,0.4,0.07,11.2,1,1,28,10
BLCK3,5.6,1.0631,-6,0.4,0.07,0.4,1,1,1,10
BLCK3,5.6,1.0631,5.6,0.4,0.07,0.4,1,1,1,10

ESEL,S,MAT,,10
NSLE,S,ALL
NSEL,R,LOC,Z,-4,4
NSEL,R,LOC,X,-4.4,4.4
ESLN,R,1,ALL
EMODIF,ALL,MAT,9
ESEL,ALL
ESEL,S,MAT,,10
EMODIF,ALL,MAT,12
ALLSEL
NUMMRG,ALL
NUMCMP,ALL

! create die
BLCK1,0.4,1.1331,3.6,0.8,7,3,0.182,0.27,0.3,3,1,2,2,8,10,2
BLCK1,0.4,1.1331,-5.2,0.8,7,3,0.182,0.27,0.3,3,1,2,2,8,10,2
BLCK1,3.6,1.1331,-2.8,0.8,3,8,0.182,0.27,0.3,3,1,2,2,8,10,2

BLCK3,0,1.1331,-3.2,3.2,0.3,6.4,8,2,16,10
BLCK3,0,1.1331,5.6,5.6,0.3,0.4,14,2,1,10
BLCK3,0,1.1331,-6,5.6,0.3,0.4,14,2,1,10
BLCK3,5.6,1.1331,-5.6,0.4,0.3,11.2,1,2,28,10
BLCK3,5.6,1.1331,-6,0.4,0.3,0.4,1,2,1,10
BLCK3,5.6,1.1331,5.6,0.4,0.3,0.4,1,2,1,10

ESEL,S,MAT,,10
NSLE,S,ALL
NSEL,R,LOC,Z,-4,4
NSEL,R,LOC,X,-4.4,4.4
ESLN,R,1,ALL
EMODIF,ALL,MAT,11
ESEL,ALL
ESEL,S,MAT,,10
EMODIF,ALL,MAT,12
ALLSEL
NUMMRG,ALL
NUMCMP,ALL

! create mold
BLCK1,0.4,1.4331,3.6,0.8,7,3,0.182,0.27,0.2137,3,1,2,2,8,12,2
BLCK1,0.4,1.4331,-5.2,0.8,7,3,0.182,0.27,0.2137,3,1,2,2,8,12,2
BLCK1,3.6,1.4331,-2.8,0.8,3,8,0.182,0.27,0.2137,3,1,2,2,8,12,2

BLCK3,0,1.4331,-3.2,3.2,0.2137,6.4,8,2,16,12
BLCK3,0,1.4331,5.6,5.6,0.2137,0.4,14,2,1,12
BLCK3,0,1.4331,-6,5.6,0.2137,0.4,14,2,1,12
BLCK3,5.6,1.4331,-5.6,0.4,0.2137,11.2,1,2,28,12
BLCK3,5.6,1.4331,-6,0.4,0.2137,0.4,1,2,1,12
BLCK3,5.6,1.4331,5.6,0.4,0.2137,0.4,1,2,1,12

! create motherboard
FR4GEO1,9,0.05,0.018,0.2,0.018,0.12,0.018,0.2,0.018,0.05
FR4GEO2,9,1,1,2,1,2,1,2,1,1
FR4GEO3,9,5,2,4,3,4,2,4,3,5
```

```
BLCK7,0.4,0,3.6,0.8,7,3,0.15,,2,,2,8,2
BLCK7,0.4,0,-5.2,0.8,7,3,0.15,,2,,2,8,2
BLCK7,3.6,0,-2.8,0.8,3,8,0.15,,2,,2,8,2

BLCK8,0,0,-3.2,3.2,6.4,8,16
BLCK8,0,0,5.6,5.6,9.6,14,6
BLCK8,0,0,-15.2,5.6,9.6,14,6
BLCK8,5.6,0,-5.6,9.6,11.2,6,28
BLCK8,5.6,0,-15.2,9.6,9.6,6,6
BLCK8,5.6,0,5.6,9.6,9.6,6,6

! set material properties
EPMAT,1,9                                                ! solder
EPMAT,2,14                                               ! copper
EPMAT,3,14                                               ! copper
EPMAT,4,11                                                  ! FR4
EPMAT,5,30                                            ! polyimide
EPMAT,6,30                                          ! solder mask
EPMAT,8,29                                             ! adhesive

! die attach
MP,EX,9,10000
MP,NUXY,9,0.35
EPMAT,11,12                                                 ! die
EPMAT,12,27                                   ! molding compound

FINISH

/SOLU
GL_1,2
GLMEC1,0,0.692,0,12,0
SUB20,30,0,0,0,1

/NUMBER,1
/PNUM,MAT,1
EPLOT

ALLSEL,ALL
SAVE

SOLVE

SAVE

/POST1
F_VONM,0,0.8,5.2,5.2,0.692,0.7864,1                  ! board side
!F_VONM,0,0.8,5.2,5.2,0.9897,1.0151,1            ! component side

! the user should take note of the output quantities
! (equivalent stress and strains for the high-stiffness case)
! e.g.: s_1, e_1

! start the low-stiffness solution
/PREP7
*GET,E_SOL,EX,1                 ! obtain Young's modulus for solder
E_SOL=E_SOL/100                      ! reduce it by a factor of 100
MP,EX,1,E_SOL               ! set property for low-stiffness solder

/SOLU
SOLVE
```

```
/POST1
F_VONM,0,0.8,5.2,5.2,0.692,0.7864,1                          ! board side
!F_VONM,0,0.8,5.2,5.2,0.9897,1.0151,1                        ! component side

! the user should take note of the output quantities
! (equivalent stress and strains for the low-stiffness case)
! e.g.:  s_01, e_01

/eof                                                        ! end of input file

! following steps cannot be included with the input file because of
! the need for specific values of equivalent stress and strains
! however the remaining steps are given in parameterized form as
! follows

! find assembly stiffness
F_E_EFF,S_1,E_1,S_01,E_01,4440184.1
! the user should take note of the output quantities
! (assembly stiffness and imposed strain)
! e.g.:  E_a, eps_imp

! 1-d simulation
F_BCLIFE,8.03e4,4.62e-4,3.3,1.12e-019,1.38e-
023,298.15,0.023,263,1.9e5,
        (continued)    8100,5.53,3.35e11,E_a,eps_imp,1,0.01338582,4,100
FINISH
EXIT
FINISH
EXIT
```

INDEX

ABOUT THE CD-ROM

The CD-ROM included with this book features computer programs that complement the material contained within the book.

The disk (CD-ROM) is distributed by Kluwer Academic Publishers with absolutely no support and no warranty from Kluwer Academic Publishers. Use or reproduction of the information on the disk for commercial gain is strictly prohibited. Kluwer Academic Publishers shall not be liable for damage in connection with, or arising out of, the furnishing, performance or use of the disk (CD-ROM).

Lightning Source UK Ltd.
Milton Keynes UK
09 October 2010

160967UK00004B/9/P